Management of

Problem
Soils

in

Arid
Ecosystems

Management of
Problem
Soils
in
Arid
Ecosystems

A. Monem Balba

CRC Press
Taylor & Francis Group
Boca Raton London New York

CRC Press is an imprint of the
Taylor & Francis Group, an **informa** business

First published 1995 by Lewis Publishers

Published 2020 by CRC Press
Taylor & Francis Group
6000 Broken Sound Parkway NW, Suite 300
Boca Raton, FL 33487-2742

First issued in paperback 2020

© 1995 Taylor & Francis Group, London, UK
CRC Press is an imprint of Taylor & Francis Group, an Informa business

No claim to original U.S. Government works

ISBN 13: 978-0-367-57971-5 (pbk)
ISBN 13: 978-0-87371-811-0 (hbk)

Visit the Taylor & Francis Web site at
http://www.taylorandfrancis.com

and the CRC Press Web site at
http://www.crcpress.com

Library of Congress Cataloging-in-Publication Data

Balba, A. Monem.
    Management of problem soils in arid ecosystems / A. Monem Balba.
        p.   cm.
    Includes bibliographical references and index.
    ISBN 0-87371-811-9 (alk. paper)
        1. Soils--Arid regions.   2. Arid regions agriculture.   3. Soil management.   I. Title.
    S592.17.A73B34 1995.
    631.4'715'4--dc20   95-19746
                                                                                          CIP

Library of Congress Card Number 95-19746

CONTENTS

Chapter 5
Water Management in Arid Ecosystems

Chapter 6
Desertification

PREFACE

The total area of major soil associations as presented in the Soil Map of the World is about 13.18 billion ha. However, as pointed out by Dudal, about half of this area is in permafrost regions. Africa contains about 3.01 billion ha, most of which suffer from drought.

In order to provide food and agricultural products for the fast-growing world population. it is necessary to improve the productivity of the cultivated land and put more land into agriculture. Soil and water conservation measures should be applied to protect these two most important natural resources from deterioration and loss and to maintain high productivity.

The UNESCO Soil Map of the World shows that the bioclimatic zones hyperarid, arid, semiarid, and subhumid account for one third of the land surface of the globe and contain about 14% of the world population. Three fourths of the area of these zones lies in the semiarid region. Because of the aridity prevailing in these zones, the vegetative cover is sporadic. Accordingly, the soil content of organic matter is low. Leaching does not take place. Thus the soil contents of soluble salts, lime, and gypsum control their chemical, physical, and fertility characteristics. Also, sand is a major soil constituent in several areas. Soils having such characteristics require special management.

Much research work has been carried out on these soils to identify their problems, reclaim them, and efficiently and economically utilize them. It is thought that a book devoted to the management aspects of these soils would help those who need to deal with them. Evidently, the theoretical and academic aspects are not the main concern of this book.

Because soil and water together make agriculture, a chapter is devoted to water management in the arid ecosystems.

Desertification is most active in arid and semiarid ecosystems. Mismanagement of the soils in these regions usually leads to desertification. Accordingly, the last chapter of this book covers the desertification problem and how to combat it.

The author has been a practicing teacher, researcher, and consultant in the field of soil and water conservation and desertification in arid and semiarid regions for an extended number of years. Actually, this is not his first publication in this field, though most of his publications are in the Arabic language. The author hopes that the present version will be as useful to the readers as his past publications.

A. Monem Balba
Alexandria, Egypt

ABOUT THE AUTHOR

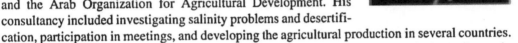

Dr. A. Monem Balba obtained his bachelor's degree in Agricultural Sciences, his diploma in Statistics from the University of Cairo, and his master's and Ph.D. degrees from the Universities of Arizona and Illinois, respectively, in Soil Science.

He worked as a soil specialist in the Ministry of Agriculture and joined the Department of Soil and Water Sciences, University of Alexandria to teach various soil chemistry and fertility courses, conduct research, and practice consultation. His main areas of research have been soil-plant quantitative relations; salt-affected, calcareous, and sandy soils; and pollution with heavy metals.

Dr. Balba served as consultant for FAO, UNDP, UNEP, and the Arab Organization for Agricultural Development. His consultancy included investigating salinity problems and desertification, participation in meetings, and developing the agricultural production in several countries.

In the last 15 years, Dr. Balba formed a group with his former graduate students. One main activity of this group is scientific publication. The author's contribution in this field varies from technical papers to reviews and books in the Arabic and English languages and publishing two scientific journals.

CHAPTER 1

The Arid Ecosystem

ELEMENTS OF THE CLIMATE

The climatic elements comprise several natural phenomena, which together constitute the climate of a particular region. Among these phenomena are the solar radiation, the atmospheric pressure, and the winds. Precipitation is a product of the interaction between the three natural phenomena and their effect on water bodies.

The Solar Radiation

When the radiation, emitted from the sun, reaches the surface of the earth, it is reflected back to the troposphere (the lower portion of the atmosphere). The reflected radiation, termed terrestrial radiation, is a factor in warming the atmosphere through its content of gases, such as carbon dioxide, and water vapor as well as dust.

The solar radiation consists of (Sharaf[1]; Howard[2]):

- The invisible thermal radiation, or infrared radiation, constitutes about 46% of the solar radiation.
- The visible radiation or sunlight rays constitutes about 45% of the solar radiation.
- The violet and ultraviolet radiation, termed by some authors "biological radiation," constitute about 9% of the solar radiation.

The solar radiation loses during its way to the earth surface about 32% by reflection with clouds and about 2% reflected to the space from the earth surface. Thus, the total reflectivity to the space amounts to about 34% of the solar radiation and is termed "Earth albedo" (Howard[2]).

The remaining 66% of the solar radiation reaches the earth's surface and is used in warming the contacting air and the atmosphere as a whole. The earth absorbs the solar radiation and transforms it into heat. At the same time, it is considered a "radiating body." The atmosphere constituents (mainly water vapor and carbon dioxide) absorb very limited amounts of the short rays of the solar radiation but have the ability to absorb much of the terrestrial radiation reflected from the earth's surface especially through carbon dioxide. This phenomenon is termed "The Greenhouse Effect."

The solar radiation is at its maximum at the equator and gradually decreases toward the two poles.

Heat is also transmitted through the air by thermal conductance. It takes place when the air having lower temperature is in contact with the earth surface with higher temperature. Heat conductance ceases when the temperatures of both the air and the earth surface are similar. Heat is also conducted from the warm air to the colder earth surface as in winter or in cold nights (Trewartha[3]).

1

When the air temperature is increased as a result of the heat conductance from the earth, the air becomes lighter and ascends. This phenomenon is termed "Convection current." The cold air in the upper part of the atmosphere is dense and becomes heavier, thus it descends to replace the warm air which had ascended aloft. Also, the latent heat of the water vapor above vast water bodies participates in warming the air (Abu El-Einein[4]).

Factors Affecting Solar Radiation Strength

The solar radiation which reaches the earth surface varies according to:

- The form and kind of the radiation waves.
- The forms of the recipient surfaces.
- Dark-color bodies absorb all radiation they receive, while the white snow reflects the high radiation it receives.

The strength of the solar radiation and its extent on the earth surface may be summarized as follows (Abu El-Einein[4]):

- Stability of the solar radiation strength termed "Solar constant" which depends on:

  a) The strength of the solar radiation.
  b) The variations in the distance between the earth and the sun according to the earth movement around the sun.

- Degree of the atmospheric clearance.
- Variations in the number of sunny hours during the day from one place to another.
- The solar radiation angle with the earth surface. The perpendicular radiation in the equatorial region crosses a shorter distance in the atmosphere as compared to the solar radiation falling on the earth surface at the polar regions. Also, the perpendicular radiation is stronger than the nonperpendicular radiation.

Any change from the average intensity and composition of the normal solar beam (at the top of the atmosphere) will produce variations of the climate. Also, climatic variations may take place upon changes in the earth's surface constituents (soil, rock, ice, water, and biota) as well as in the gaseous and particulate composition of the atmosphere (Hare[5]).

The vegetative cover plays a role in the radiation balance. Accordingly, overgrazing reduces this balance. In order to maintain thermal equilibrium, and in the absence of advection, the air must descend and compress adiabatically (causing adiabatic warming), thus decreasing the relative humidity and precipitation (Charney[6]).

The Atmospheric Pressure

The atmospheric pressure on any point of the earth's surface is the weight of a column of air above this point extending to the higher ends of the atmosphere. This weight above an area of 1 cm^2 at the sea level is equal to the column of mercury having a cross section of 1 cm^2 and a length of 76 cm. It was also known that this weight—the atmospheric pressure—varies with time for any particular site and at various locations at the same period of time.

The air temperature is an important factor in the atmospheric pressure variation. A negative correlation exists between the atmospheric pressure and the air temperature. Thus, when the air temperature rises, the air size expands and its density decreases causing a decrease in air weight and its pressure, and vice versa.

The Horizontal Pressure Distribution (Abu El-Einein[4])

The factors which affect the horizontal distribution of air temperature at the earth surface are the same factors affecting the atmospheric pressure at one place or the other. In addition, the location of a place relative to the tropical region and the variations in the geographic distribution of land and water have considerable effect on the horizontal variability of the atmospheric pressure on the earth surface. The range of distance between one place and the tropical zone results in differences in its temperature and pressure from other places. The following pressure distribution zones almost coincide with temperature distribution zones:

Equatorial Low

This zone extends between latitude 5° north and south. The atmospheric pressure is as low as 1013.2 mbar. The wind flows to this zone from the tropical zone.

Subtropic Highs

This zone comprises the two hemispheres of the tropic region north and south of the equator between 25–30° north and south. The wind flows from this zone to the equatorial zone. This region is traversed twice by the highest pressure and is likely to be dominated by subsidence skies and rainlessness. Several factors interfere to divert this direction:

1) Variations in the solar radiation. The solar radiation is at its maximum in the region between the two tropics and decreases beyond this tropical region. It reaches about half its magnitude at each pole as compared with the tropical region. Thus, the thermal radiation accumulates in the tropical region. The absorbed radiants exceed the reflected resulting in a positive thermal balance in the tropical region, contrary to the other regions. A permanent wind movement is thus initiated at the earth surface from the high atmospheric pressure centers which are low in temperature to the low atmospheric pressure centers which have high temperatures. This movement takes place to reestablish the thermal balance in the troposphere layer in general and in the air contacting the earth surface in particular. Thus, the wind movement close to the earth surface is related to the permanent and seasonal atmospheric pressure centers which are affected with variations of the temperature of the air contacting the earth surface during the seasons of the year. These factors are termed thermal factors (Abu El-Einein[4]; Howard[2]).
2) The deflective force resulting from the oscillation of the earth around its axis (Coriolis force).
3) The frictional force resulting from the stress of the wind at the earth's surface. This force decreases the wind speed.

4) The physiography of the earth's surface such as the high mountains which decrease the rate of flow of the wind near the earth's surface.
5) Diurnal variations of wind speed also take place. Thus, the maximum wind speed is at noon (middle of the day, intensive radiation and lowest atmospheric pressure) and the minimum is just before sunrise (lowest temperature and highest atmospheric pressure).

The atmospheric pressure is affected by the temperature of the air. At the same time, the atmospheric pressure and gradient affect the wind direction and speed on the earth's surface. Since solar radiation in the equatorial and tropical regions is positive, the gained heat surpasses the lost heat. Thus considerable latent heat, present in these regions, uplifts the warm air saturated with water vapor. The warm air flows upward, especially above the large water bodies forming ascending air currents. When the ascending air reaches saturation with the water vapor, it starts to condensate forming clouds, snow, hails, or rains. Hence, uplifting of humid air is a basic step for the formation of rains. When air subsides, it warms up and, accordingly, its capacity to hold moisture increases, resulting in inhibition of rain formation. Under such conditions, aridity prevails.

Rain does not fall unless stability is disturbed enough to cause air uplift. A combination of factors keeps the atmosphere stable for long periods in most parts of the dry regions. When temperature decreases slowly with height, the atmosphere is stable. Stability is favored if the moist lower layers of the atmosphere are warmed aloft or cooled below. This happens when: (1) the air moves downwards which leads to dynamical warming aloft, and (2) warm air passes over cold surfaces.

The combination of the high soil temperature induced by bright sunrise and dryness of the atmosphere, produces strong long-wave radiative cooling of the surface. Thus, the surface tends to be light in color and to have a high albedo. The net radiation is, accordingly, low due to the reflection (Hare[5]). Hare summarizes the main causes of aridity as follows:

• widespread, persistent atmospheric subsidence.
• localized subsidence induced by mountain barriers or other physiographic features.
• absence of rain-inducing disturbances causes dry weather even in areas of moist air.
• absence of humid air streams.

Dry climates are characterized by their variability. To measure this variability (v), the coefficient of variation is used (V_Q)

$$V_Q = \frac{Q}{P} \times 100 \qquad (1)$$

where Q is the standard deviation of annual precipitation and P is the mean annual precipitation.

In most arid regions V_Q is over 25%. It exceeds 40% along most desert margins.

Wallen[7] and Wallen and Perrin de Brichambault[8] suggested instead of the coefficient of variation to use the percentage ratio of the mean interannual precipitation difference ($P_{n-1} - P_n$) to the mean annual precipitation P (N − 1)

$$V_1 = 100 \; / \; \frac{(P_{n-1} - P_n)}{P \, (N - 1)} \tag{2}$$

where n is an individual year in a series of N years.

This ratio for the Middle East was found over 50% along the desert margin. In dry farming areas, it ranged between 25 and 35%. When P is in millimeters, they[7,8] suggested the following empirical equation along the dry margin of arable farming

$$V_1 = 0.007 \; P + 2.2 \tag{3}$$

Water Evaporation

Precipitation is not a single parameter for characterizing the climate of a region. It is the effective precipitation that participates in the various activities in the soil. Several investigators utilize the ratio between precipitation and evaporation of water in a region to characterize the region's climate.

Evaporation is a physical process which leads to the loss of water from the soil or water surfaces to the atmosphere. Kuenen[9] and Panal[10] reported that the rate of water evaporation depends on several factors among which are:

1. Temperature of the water and the air.
2. Relative humidity of the air.
3. Supply of heat needed for evaporation.
4. Rate at which adjacent air is replaced.
5. Salinity of the water (Balba and Soliman[11]).

In this regard, dissolved inorganic salts in water change many of the water properties such as boiling point, freezing point, and entropy changes. The magnitude of these changes is dependent on salt concentration and salt kind.

The following cases characterize water evaporation from bare soil (Rijtema[12]):

1) When the soil water table is reasonably close to the soil surface, the evaporation from the soil "E_s" is not limited by water transmitting properties of the soil, or $E_s < V_{max}$, where "V_{max}" is the maximum flux rate by capillarity. In this case, the flux through the soil is controlled completely by meteorological conditions.
2) When $E_s = V_{max}$, the flux through the soil surface is just sufficient to maintain evaporation rate "E_s." The difference from the first case is different suction or moisture distribution in the soil profile above the water table.
3) When $E_s > V_{max}$, the water transmitting properties of the soil are limiting the evaporation.

Qayyum and Kemper[13] showed that when saline water moved upward, salts tended to accumulate near or at the soil surface, causing a reduction in the rate of supplying water to the soil surface than the evaporation rate. A dry surface layer was formed which acts as a barrier to the movement of liquid and vapor water. Thus, a reduction in evaporation rate takes place.

Evapotranspiration

The consumptive use or evapotranspiration is the sum of two terms:

a) Evaporation which is the process that leads to the loss of water from the soil or water surfaces to the atmosphere as stated above.
b) Transpiration which signifies the process by which water vapor escapes from the leaves and enters the atmosphere.

Transpiration by plant leaves has received considerable investigation. Factors that affect the rate of transpiration are:

A) Plant factors

- Leaf area, Kramer.[14]
- Leaf structure: Cutinized leaves transpire less than uncutinized leaves even when the cutinized leaves are larger in surface, Turrel.[15] Also, size and form of leaves are less-than-important factors, Stalfelt.[16]
- Root-shoot ratio: Transpiration per unit leaf area increased as the ratio of root surface to leaf surface increased, Parker.[17]
- Water content of leaves, Ketellaper.[18]
- Sunken stomata: surface, Meyer and Anderson.[19]

B) Environmental factors

- Solar radiation, Robbins et al.[20]
- Relative humidity, Hill et al.,[21] Gates and Hanks,[22] Oleary and Knocht.[23]
- Air temperature, Gates.[24]
- Wind velocity, Robbins et al.[20]
- Atmospheric pressure.
- Soil conditions influencing water availability.

C) Effect of salts on root growth and water absorption, Balba and Soliman.[25]

Estimation of Potential Evapotranspiration (ET)

The determination of ET is a complex problem shared by meteorology, soil, and plant sciences. Because of the difficulty of direct measurement, a number of empirical and semiempirical equations have been proposed for estimating ET using readily available climate data: monthly air temperature, relative humidity such as mean sunshine percentage, solar radiation, etc. Among these equations are the Penman,[26] Blaney-Criddle,[27] Hargreaves,[28] Rijtema,[29] and Von Bavel.[30]

The Penman Equation

Penman[26] introduced his equation in 1948. It combines aerodynamic and energy balance approach. Although it is more complicated than other formulas, it is widely used for the determination of potential ET.

The Penman equation appears as follows:

$$ET = \frac{DH + 0.27\ E_a}{D + 0.27} \tag{4}$$

where   H   =   RA $(1 - r)$ $(0.18 + 0.55\ [n/N]) - QT_a^4\ (0.56 - 0.92\ ed)\ (0.10+0.90$
              $[n/N])$

   E_a = $0.35\ (e_a - e_d)\ (1 + 0.0098U_2)$
   H   =   net incoming radiation, mm H_2O/day
   RA   =   mean monthly extra terrestrial radiation, mm H_2O/day
   r   =   reflection coefficient of the surface
   n/N   =   sunshine percentage
   Q   =   Boltzman constant
   e_d =   actual vapor pressure in the air, mm Hg
   e_a =   saturation vapor pressure at main air temperature, mm Hg
   E_a =   evaporation, mm H_2O/day
   U_2 =   mean wind speed at 2 m above the ground, mile/day
   ET   =   evapotranspiration, mm H_2O/day
   D   =   slope of saturated vapor pressure curve of air at absolute tempera-
              ture T_a in $F°$, mm Hg/$F°$
   0.27   =   psychrometer constant.

Penman's equation requires that the crop must: (a) never be short of water, (b) have uniform height, (c) be green, (d) be actively growing, and (e) be completely shading the ground throughout the whole of its growing season.

CLASSIFICATION OF THE EARTH TO CLIMATIC REGIONS

The above-stated climate elements are basic forces affecting the climate of a region through their effect on several natural phenomena: air temperature, rainfall, relative humidity, dust and sand storms, evaporation, dew, and light. These phenomena control the plant species that may grow in any phytogeographic area.

According to Meigs[31] one of the simplest classifications stated that 250 mm of annual rainfall is the dividing line between arid and semiarid and 500 mm of rainfall between semiarid and humid.

Coe[32] stated that Harris[33] used the annual rainfall to classify the arid regions. According to Harris, the semiarid regions lie below 300 mm of annual precipitation and the semihumid savanna lies between 500 and 1000 mm.

Most present classifications use combinations of temperature and precipitation. De Martonne[34] used the annual precipitation and temperature. His formula appears as follows:

$$\text{The index of aridity I} = P\ /\ (T + 10) \tag{5}$$

where "P" is the mean annual rainfall and "T" is the mean monthly temperature.

Koppen[35] suggested a margin between arid and semiarid climates: R (rainfall) = T + 11. He suggested formulas to assign climatic values to the limits of the main vegetation types of the world.

Thornthwaite[36] followed Koppen[35] in considering that the natural vegetation reflects the collective effect of the climate elements in any specified region. Accordingly, Thornthwaite[36] suggested borders between the climatic regions depending on differences in soil characteristics, natural vegetation, and natural drainage.

The precipitation effectiveness required for plant growth can be calculated by dividing the accumulated rain evaporation "E," i.e., "P/E." Using per month "P" by the total, the total "P" and "E" per year gives the precipitation evaporation index.

Thornthwaite[36] distinguished five moisture classes on the basis of the "P/E" index.

Moisture class	Natural vegetation	P/E index
A - Wet	Very dense forest	128
B - Humid	Forests	64–127
C - Subhumid	Savanna	32–63
D - Semiarid	Steppe	16–31
E - Arid	Desert Grasses	16

Thornthwaite[36] gave to rain seasonality more importance than total rain values. Thus, he added four letters to each moisture class suggested above indicating the rainy season and its intensity "Seasonal concentration of rainfall" as follows:

R = rainfall year round.
S = rainfall decreases in summer.
W = rainfall decreases in winter.
d = rainfall decreases year round.

Based on the thermal efficiency or "Temperature/evaporation" T/E index, Thornthwaite[36] distinguished the following thermal regions:

Thermal regions	T/E index
A - Tropical	> 128
B - Mesothermal	64–127
C - Microthermal	32–63
D - Taiyal	16–31
E - Tundral	1–15
F - Frost	Zero

Taking the above three bases for distinguishing and classifying the earth to climatic classes one can end up with about 120 climatic regions.

In 1948 Thornthwaite[36] used the above-stated three bases, namely the moisture class according to P/E index, the seasonal and concentration of rainfall, and thermal region according to T/E, to modify the limits for the climatic regions. He emphasized in this classification the importance of potential evapotranspiration because it affects the relative humidity in the atmosphere.

Emberger[37] used mean annual precipitation "P," mean daily maximum temperature of the warmest month "M," and mean daily minimum temperature of the coldest month "m," all combined into a single moisture quotient "Q" using the following formula:

$$Q = 2P / (M + m) (M - m) \times 100 \qquad (6)$$

Budyko's[38] classification of arid and humid regions is widely used at present. It depends on the "moisture index" (IM) of Thornthwaite,[36] defined as:

$$IM = 100 (P / EP - 1) \qquad (7)$$

where "EP" is the potential evapotranspiration.

Budyko[38] introduced the "radiation index of dryness" (D), defined as:

$$D = 1 - C / S \tag{8}$$

where C   is the runoff ratio which is the water surplus divided by the precipitation
      S   is Priestly-Taylor parameter and
          is parameter related to the rate of variation of saturation vapor pressure with
          respect to temperature as shown by Hare.[5]

The Budyko[38] equation thus appears as follows:

$$\text{The dryness ratio is } D = R / LP \tag{9}$$

where R   =   mean annual net radiation, i.e., radiation balance
      P   =   mean annual precipitation
      L   =   latent heat of vaporization for water.

In hot dry climates, energy supply is excessive and rainfall is deficient. Thus, the index of aridity, the dryness ratio, would depend on their relative magnitudes, and the dryness ratio at a given location indicates the number of times the net radiation energy income of the surface could evaporate the mean annual rainfall.

According to Budyko[38] the arid regions are divided into the following according to the value of "D":

Dryness "D"	Vegetational response
> 3.4	Desert
2.3–3.4	Semidesert
1.1–2.3	Steppe or Savanna

Gresswell[39] described four main climates as follows:

1. Equatorial
2. Tropical
3. Temperate
4. Arctic

As long as we are dealing with the arid and semiarid ecosystems, which fall in Gresswell's[39] "Tropical climates," a somewhat detailed description of these climates will be stated (Abu El-Einein[4]).

The tropical climates include the following:

1. The humid semitropical climate which does not concern the present discussion.
2. The tropical and semitropical grassland climate.
3. The dry hot desert climate.

The Tropical and Semitropical Grassland Climate

The semiarid ecosystem falls in this region. This climatic region is affected by the continental tropical warm dry air masses. It is bordered in the north by the desert dry hot region and in the south by the humid dry region, in the north portion of the earth. This is especially clear in Africa. South of the equator, this region lies between Benguola in the west and Bolawayo in the east. It is also present in the United States, South America, and north of Australia.

In this climatic region, mean annual temperature is about 20°C, the annual thermal range is about 8–15°C. About 600 mm of rain falls annually, 60% of which falls in summer.

Rainfall in the arid, semiarid, and subhumid regions is highly erratic and takes place in heavy brief unevenly spaced showers during the rainy seasons. In these regions, the role of soil moisture and the vegetative cover is not negligible in water and heat balance of the atmosphere. The evaporation over the dry surfaces is reduced and subsequently precipitation and release of latent heat may be reduced.

The Hot Dry Desert Climate

This region is the main source of the continental dry tropical air masses. Accordingly, it represents the areas of high pressure center and the subsiding air. The air content of humidity is low. The African Great Desert, "the Sahara," represents a main region of this climate. It extends eastward from the Atlantic Ocean to the middle of Asia.

The temperature of this climatic region is not affected by the distance to the equator but by its spacious land area, the cloudless sky, and kind of marine currents at the coasts. Temperature rises by day during the summer season and considerably decreases at night and in winter. Precipitation is rare and hardly surpasses 200 mm annually.

Generally, this region is characterized by the subsiding air, adiabatic warming, low relative humidity, high potential evapotranspiration, and rare precipitation. Accordingly, it is considered very arid, having low soil moisture and sporadic vegetative cover.

In the opinion of Harris,[33] the length of the dry season is an important factor in the arid lands' utilization. He pointed out that the annual variations in rainfall constitute an important problem. In the semiarid savanna, rain is expected to vary from one year to another by 25–40%. He also noted that short sharp showers usually occur where the degree of aridity increases. Under such conditions precipitation is likely to exceed the infiltration capacity of the soil and its effectiveness for plant growth is decreased while soil erosion may increase.

In preparing "World Map of Desertification"[40] by UNEP, UNESCO, FAO, and WMO, a bioclimatic map was used to show zones of aridity. The limits of the zones were decided primarily from a climatic aridity index:

$$P / ET = \text{Precipitation} / \text{Evapotranspiration}$$

The potential evapotranspiration is calculated by the method of Penman,[26] taking into account atmospheric humidity, wind, and solar radiation.

1. The hyperarid zone (P/ET < 0.03). This zone corresponds to the extreme desert.
2. The arid zone (0.03 < P/ET < 0.20) comprises dry land areas with sparse perennial annual vegetation. Nomadic pastoralism can be practiced here, but rain-fed agriculture is not possible.
3. The semiarid zone (0.20 < P/ET < 0.50) includes steppe or tropical shrub and with discontinuous herbaceous layer and increased frequency of perennials. Livestock breeding and rain-fed agriculture are both possible in this zone.

4. The subhumid zone (0.50 < P/ET < 0.70) is characterized by more dense vegetation. It includes tropical savannas, mediterranean maquis and chaparral, and steppes with chernozem soils. Rain-fed agriculture is common in this zone for crops adapted to seasonal drought.

The area of the four bioclimatic zones (hyperarid, arid, semiarid, and subhumid) of the UNESCO map[40] accounts for approximately one third of the land surface of the globe and contains approximately 14% of the world population, almost 3/4 of which is concentrated in the belt classified by UNESCO as semiarid (Man and Biosphere[41]).

According to Kassas,[42] aridity is a "situation" in which water income is less than potential water expenditure. He distinguishes between the following deserts:

a) Rainless deserts, where rainfall is not an annually recurring event.
b) Runoff deserts, where rainfall is less than 100 mm and variable and where the perennial plants are restricted to specially favored habitats.
c) Rainfall desert, where rainfall is insufficient for sustained crop production (100–200 mm) and where perennial plant life may be widespread and not confined to runoff collecting habitats.
d) Man-made deserts due to "desertification."

NATURAL VEGETATIVE COVER IN ARID ECOSYSTEMS

Because of spotted precipitation and due to variability of surface topography, the vegetative cover in arid regions is sporadic. It consists of ephemeral and perennial plants.

Desert Ephemerals

Desert ephemerals develop in the rainy season. Among the dominant species are *Mesembryanthemum forskalea, Zygophyllum simplex, Trigonella stellatta, Erodium bryoniafolium, Schismus calcynus,* and others.

The roots of these plants absorb their need of water and nutrients from the surface layer of the soil. Ephemerals do not economize water but they escape drought by completing their life-cycle before the dry season. They are considered by Ayyad and Kamal[43] as "drought-evaders."

Desert Perennials

Desert perennials dominate in the dry seasons and include *Haloxylon salicornicum, Retama raetam, Citrulus cocynthis, Hoyssymus muticus,* and others. According to Migahid[44] the perennials have the ability to absorb water from deeper permanently wet soil layers. They are considered by Ayyad and Kamal[43] as "drought enduring."

The desert perennials acquire properties to increase their ability to absorb soil water and/or decrease transpiration and regulate the intensity of water loss. There is a number of measures the desert perennials apply to endure the arid conditions in their habitats. Similar properties characterize the suitable crops for sandy soils which suffer water stress.

Extensive Roots

Because of the open structure of the vegetation due to the sporadic plant distribution, the desert perennials' roots can extend freely in every direction. In general, the depth and breadth of the roots are several times as great as the length and lateral extension of the shoots. Vigorous rooting is generally a property associated with drought tolerance (Gul and Allan[45]).

Xeromorphic Structure

Perennials of the desert develop xeromorphic structure by which they are able to decrease transpiration. Migahid[44] stated the following examples:

1. Thick cuticle and waxy coating. Wright and Dobrenz[46] showed that water use efficiency and seedling drought tolerance are associated with the quality of cuticular wax on the leaves of Lehman lovegrass.
2. Sunken stomata.
3. Spine formation.

Reduction of Transpiring Surface

Desert plants usually bear tiny leaves or may be leafless to minimize the transpiring surface. Miskin and Rasmusson[47] reported that the number of stomata per unit leaf area influenced transpiration and stomatal diffusive resistance.

Low Tissue Water Content

Water content of xerophytes is generally lower than that of mesophytes, though plant water content may be increased by special mechanisms such as succulence and osmotic pressure adjustment.

Binding of Water

The proportion of bound to free water in the plants increases when the plants are accustomed to growing in a dry habitat.

Succulence

By succulence it is meant a high ratio of plant water content to the plant dry weight or the high ratio of the water content to leaf surface. Many xerophytes and mesophytes are succulent such as *Hesembryanthemum forskalea, Zygophyllum simples,* and others. Chapman[48] as cited by Ayyad and Kamal[43] reported that the phenomenon of succulence is generally common in the vegetation of saline areas as it is a mechanism which enables plants to overcome dryness caused by high osmotic pressures of the root environment. Balba and Soliman[25] showed that Sudangrass plants grown in a saline habitat contained more water than the same plants grown on a nonsaline medium. Increase of a plant's water content was associated with lower transpiration (see Chapter 2, Table 7).

High Osmotic Pressure

The mechanisms of water uptake by plants and its loss by evaporation from the soil and by transpiration from plants are controlled by various processes among which is the osmotic pressure of the soil solution and/or cell sap.

1. Soil solutions with high osmotic pressures depress their rate of evaporation from the soil as shown by Balba and Soliman.[25]
2. Transpiration of the cell sap having high osmotic pressure is lower than cell sap with low osmotic pressure (Migahid[44]).
3. High osmotic pressure causes lowering of the permeability of the transpiring cells to water.
4. Plants having high osmotic pressures for their cell sap are more capable to absorb water.

Because of this important role of the cell sap high osmotic pressure, plants adapted to arid ecosystems usually acquire this important property. Plants which face water stress in periods of their growth may acquire the ability to adjust the osmotic pressure of their cell sap to be able to endure the water stress to maintain the leaves' turgor. Walter,[49] Repp,[50] and Begg and Turner[51] considered it the most effective mechanism which plants apply to tolerate or avoid situations of less water availability due to drought or salinity. However, there are crops that do not display adjustment of the osmotic pressure of their cell sap as stated by Turk and Hall.[52]

According to Radin,[53] adjustment of osmotic pressure to counteract difficulties to absorb water due to excess salinity is different from counteracting difficulties imposed by water difficulty:

a. Under saline soil solutions, plants can use the soil solution salt content to raise their cell sap osmotic pressure. Under drought conditions osmotic adjustment results largely from "internally-generated osmotica."
b. The extension of the root zone as well as the other means stated above (Migahid[44]) minimize the effect of drought on the plants.

SOILS OF THE ARID ECOSYSTEMS

The soils of arid regions are the product of several factors, mainly climate, parent rocks, and topography. Other factors of soil formation processes are slow and physical in nature. The differences between day and night temperatures cause disintegration of rocks. The strong winds and hot dry climate combine to encourage blowing of the fine particles. Transportation of fine materials prevails until the pebbles and small stones accumulating on the surface form what is called "desert pavement." The rain is scanty and only slow gentle rain can penetrate such paved surfaces. Thus, chemical reactions that make the soil constituents and horizons are not active and little or no leaching takes place. When rain falls, it takes the form of short sharp showers which usually cause severe erosion.

The dominant rocks are limestone, sandstone, and metamorphic rocks. Wind-blown sand constitutes a major part of the soil material. Gypsum is encountered in variable depths. Salts accumulate as leaching is not active.

Topography plays an important role. Areas surrounded by elevated hills receive runoff water and deposits more than other areas.

Because of the arid climatic conditions, vegetation is sporadic and accordingly, the effect of vegetation on soil formation is not pronounced. Also, dry warm climatic conditions enhance the decomposition of plant residues, thus further weakening the effect of the organic matter on the soil characteristics. The activities of humans under the above described conditions do not significantly contribute to soil formation unless irrigation is introduced or in wadis where agricultural activities depend on runoff water.

The soil profiles under such conditions are not well developed. The factors which may be taken into account in grouping the soils for agricultural use are:

1. The depth of profile to hard rock or water table.
2. Characteristics of profile layers including texture, color, and stoniness.
3. Salinity of the saturated extract.
4. Presence and depth of impermeable pans.
5. Presence and depth of lose sand.

The prevailing soils in these regions are the Aridisols Soil Survey Staff. They have Ochric epipedon and one or more additional diagnostic horizons which may be cambic, argillic, natric, calcic, gypsic, or salic (Soil Survey Staff[54]).

Salts have a chance to accumulate under arid conditions resulting in salt-affected soils. Also, because of the same conditions, soils developed from calcic rocks preserve their content of lime. The $CaCO_3$ content of the highly calcareous soils imposes several characteristics on these soils which require specific management. Wind-blown sands cover vast areas. Agricultural utilization of these sandy areas also requires special packages of technologies.

Classification of Soils of Arid Ecosystems

According to Soil Survey Staff,[54] the order Aridisols contains two suborders:

1. Argids which have an argillic or natric horizon and other Aridisols; Orthids. The Argids contain several great soil groups such as:

   a. Durargids, characterized by the presence of a hardpan below the argillic clayey horizon.
   b. Nadurargids, the hardpan is below a natric horizon.
   c. Natrargids, presence of a natric horizon without a petrocalcic horizon.
   d. Paleargids.
   e. Haplargids.

2. The Orthids contain several great soil groups such as:

   a. Salorthids which have a salic horizon whose upper boundary is within 73 cm of the soil surface and are saturated with water within a depth of 1 m for one month or more in most years, or have artificial drainage and lack duripan that has an upper boundary within 1 m of the soil surface.

b. Palorthids are soil that have a petrocalcic horizon whose upper boundary is within 1 m of the soil surface and is not overlain by a duripan.

c. Durothids, soils that have a duripan whose upper boundary is within 1 m of the soil surface.

d. Gyporthids, soils that have a gypsic or petrogypsic horizon whose upper boundary is within 1 m of the soil surface.

e. Calciorthids, soils that have a calcic horizon whose upper boundary is within 1 m of the surface and that are calcareous in all parts above the calcic horizon. Below the upper soil to a depth of 18 cm is mixed unless the texture is as coarse or coarser than loamy fine sand.

f. Camborthids, altered soils that have cambic altered horizon.

From the arid to semiarid ecosystems, temperature tends to be lower, rainfall is more frequent, humidity increases, and evapotranspiration decreases. Accordingly, plants become closer and taller. Thus, soil organic matter increases giving the soil a brownish color.

Water penetrates the soil, thus chemical reactions which form soil clay, humus, and other constituents are active. Leaching also becomes effective and soil horizons are more developed than soils of the arid zone and soil salts do not accumulate on the surface. Under such conditions, man's activities are numerous: grazing, raising animals, and growing crops in the rainy seasons. The prevailing soils in these regions are Aridisols, Alfisols, and to a lesser extent, Entisols. Oxisols and Vertisols also cover vast areas, especially in Africa (Kampen and Burford[55]). Most of these soils except Aridisols do not suffer from aridity. On the contrary, the Alfisols usually contain sufficient soil moisture. However, in dry seasons moisture stress is a dominant factor. Because most of these soils, except the Vertisols and Mollisols, are coarse textured, even the deep phases rarely store sufficient water within their profile.

The soils of the semiarid tropics are subject to high rainfall intensity. When the rains fall at the beginning of the rainy season with the soil still barren or cultivated, but before there is appreciable plant cover, severe soil erosion takes place.

Models of air cycles show that subtropical regions can be regions of subsiding air. The temperature of the air adjacent to the ground surface increases, thus its ability to hold moisture increases and rain does not form. These conditions take place in the region between latitudes 15° north and 30° south, though arid climate extends to other latitudes and other factors may interfere.

There must always have been deserts on the earth, for they are due to atmospheric subsidence, and on an unequally heated rotating earth such subsidence must always occur in subtropical latitudes (Hare[5]).

Regions of arid lands extend across the globe more than the barren deserts. Rainfall in these regions is not satisfactory for growing crops, though it suffices for grasses. Out of these regions, semiarid lands extend according to temperature (Table 1).

Drought-resistant crops may be grain accompanied with conservation measures. These areas border the subhumid region. Rainfall in semiarid regions may be as much as 650 mm/yr. It has been stated above that climatic classification of the globe does not depend solely on rainfall, as rainfall is subject to evaporation. The evaporation rate is high near the equator and during hot seasons.

The borders of the arid regions do not seem to be stable. A considerable portion of the Sahel area—south of Great Sahara in Africa—has been changing into arid land for thousands of years. The area of Lake Chad was much larger than it is now, thousands of years in the past. The Rajastan Desert was some 1500 km east of the arid region 8000 yr before. Generally, the borders of the arid regions are subject to short period changes depending

Table 1.   Distribution of salt-affected areas in developing countries (areas in 1000 ha).[1]

Country	Area	Country	Area
Mexico and Central America		**Africa** (continued)	
Cuba	316	Rhodesia	26
Mexico	1649	Senegal	756
	1965	Sierra Leone	307
		Somalia	5602
South America		Sudan	4874
Argentina	85612	Tanzania	3537
Bolivia	5949	Tunisia	990
Brazil	4503	Zaire	53
Chile	8642	Zambia	863
Colombia	907		80529
Ecuador	387		
Paraguay	21902	**Southeast Asia**	
Peru	21	Indonesia	13213
Venezuela	1240	Cambodia	1291
	129163	Malaysia	3040
		Sarawak	1538
Africa		Thailand	1456
Afras & Issas Territory	1741	Vietnam	983
Algeria	3150		21521
Angola	526		
Botswana	5679	**South and West Asia**	
Chad	8267	Afghanistan	3101
Cameroon	671	Bangladesh	3017
Egypt	7360	Burma	634
Ethiopia	11033	India	23796
Gambia	150	Iran	27085
Ghana	318	Iraq	6726
Guinea	525	Israel	28
Guinea Bissau	194	Jordan	180
Kenya	4858	Kuwait	209
Liberia	406	Muscat and Oman	290
Libya	2457	Pakistan	10456
Madagascar	1324	Qatar	225
Mali	2770	Saudi Arabia	6002
Mauritania	640	Sri Lanka	200
Morocco	1148	Syria	532
Namibia	2313	United Arab Emirates	1089
Niger	1489		83570
Nigeria	6502		

Source: Massoud.[18]

on rainfall fluctuation. These climatic changes are shown geographically by expanding or shrinking the arid areas resulting in the change of semiarid regions to arid or subhumid regions (UNCOD Secretariat[57]; Goudie[58]).

The desert environment is characterized with:

a) high thermal stress on the inhabitants as high ambient air temperature in summer coupled with high intensity of solar radiation results in appreciable heat stress.

b) The energy inflow from the sun into the arid zones is very high and much of it penetrates to the ground surface due to clear skies prevailing most of the year. Biomass potential production from the energy point of view is very high provided water is satisfactory.

c) The selective rechanneling of energy into biomass has been constrained by the shortage of water. Not only the annual total precipitation is small, but also still a considerable loss of water through evaporation, seepage to untapped aquifers, and through uncontrolled runoff take place. However, there remains considerable potential for the development and better management of water resources.

Rain-fed agriculture and pastoralism can be adjusted to the risks associated with droughts.

Other sources of water* in addition to precipitation are groundwater and major rivers. Both sources have been supplying inhabitants of the arid regions with copious amounts of water (UNCOD Secretariat[57]; Goudie[58]).

Factors Affecting the Fertility of Arid Soils

It has been stated above that the soils of arid and semiarid regions are poor in organic matter. Accordingly, soils are deficient in nitrogen. Because leaching hardly takes place under arid climatic conditions, these soils are usually saturated with cations and have basic pH values. The soil cation exchange capacity depends on the soil texture, the prevailing type of clay and organic matter. The exchangeable cations are considered available to plants. Thus, these soils are generally rich in cationic nutrients. Under basic pH, phosphorus and micronutrients are present as precipitates. Sands may be a significant portion of the soil constituents. The sandy soils are known to be low in their fertility (Balba[59]).

Geographical Extent

The arid ecosystems prevail in vast areas. They include the deserts and the arid lands that receive 100–200 mm of rainfall. Based on climatic data, the area of the world's deserts constitutes 36.8% of the earth's land surface. Whereas, according to Schantz,[56] on the basis of soil and vegetation data the total area of the world's deserts equals 43% of the earth's land surface.

There are five main desert belts:

1. A vast belt extending from the Atlantic Ocean to China including the North African Desert, the desert area in the eastern Arab countries (Syria, Jordan, Saudi Arabia, and Iraq), Iran and Middle Asia, the Russian Common Wealth countries and Afghanistan deserts, Rajastan desert of Pakistan and India, and Takla Makan and Goba deserts in China.

2. The Kalahari Desert of southern Africa and much of the plateaus of the interior.

3. Most of the continent of Australia.

---

* Water resources and management are further discussed in Chapter 5.

4. The Sonoran Desert of northwestern Mexico which extends to the southwestern United States.

5. A narrow strip of coastal South America from the equator to 35° S west of the Andean slopes, plus a broader strip east of the Andes from 18° S to southern Patagonia, together with small areas in eastern Brazil, Colombia, and Venezuela.

More extensive than the extreme deserts are the world's arid lands without enough rain to support cropping but with sufficient vegetation to support pastoralism (Table 1).

The arid lands of the world fall within the territories of over 110 national governments and the homes of roughly 600 M people.

As one passes from the true desert toward humid regions several changes take place. Rainfall increases and the mean annual rainfall tends to be constant. The vegetative xerophytic cover yields to open savannah woodland toward deciduous and evergreen tropical forest.

With the change in climate and vegetation, the land use pattern is changed. Semipermanent pastoralism and agriculture replace nomadic pastoralism.

The dryland ecosystems are able to maintain balanced exchanges of water and energy provided vegetative cover is not removed. Exposing these lands to high temperatures and erosive forces result in mineralization of organic compounds, the leaching of soil minerals, and destruction of the soil structure. Erosion may remove the entire surface. Also, water infiltration through the soil decreases which may affect the groundwater level. Several other problems are initiated as a result of utilizing soil of the arid and semiarid ecosystems.

REFERENCES

1. Sharaf, A.A.T., Climatic and Botanic Geography (In Arabic), Part 1, 3rd Ed., p. 36, Alexandria, 1961.
2. Howard, J.C., General climatology, 2nd Ed., p. 14, 1966, Cited in Abu El-Einein.[4]
3. Trewartha, G.T., An Introduction to Climates, p. 20, McGraw-Hill, New York, 1954.
4. Abu El-Einein, H.S. (In Arabic). Basis of Climatic Geography. Aldar Al Gameiyah, Alexandria, pp. 545, 1981.
5. Hare, E.K., Climate and Desertification. In Desertification, Its Causes and Consequences. Secretariat of UN Conference on Desertification, Nairobi, Kenya, Pergamon Press, New York, 1977. Chapter 2.
6. Charney, J., Dynamics of Desert and Drought in the Sahel 1975, Cited in Hare.[5]
7. Wallen, C.C., Aridity definitions and their applicability, Geografisca Amaler 49 A, 367, Cited in Hare,[5] 1977.
8. Wallen, C.C. and Perrin de Brichambault, A., Study of agroclimatology of semiarid and arid zones of the Near East. FAO / UNESCO / WMO Interagency Project on Agroclimatology, 1962, Cited in Hare,[5] p. 87.
9. Kuenen, P.H., Realms of Water, Some Aspects of its Cycle in Nature, p. 91, John Wiley & Sons, New York, 1955.
10. Panal, A., Estimating evapotranspiration. An evaluation of techniques. Australian Water Resources Conf., Hydrology Series No. 5, 1970.
11. Balba, A.M. and M.F. Soliman, Effect of kind and concentration of solute on water evaporation and salt distribution in sand columns. Alex. J. Agric. Res. 26:739, 1978.
12. Rijtema, P.W., Derived meteorological data, Transpiration, Proc. Symp. Agric. Method, Reading, p. 55, 1969.
13. Qayyum, M.A. and W.D. Kemper, Salt concentration gradients in soil and their effects on moisture movement and evaporation. Soil Sci., 93:333, 1962.

14. Kramer, P.J., Transpiration and the water economy of plants. In Plant Physiology, Stewart, F.G., Chapter 27, p. 625, 1959.
15. Turrel, F.M., The area of internal exposed surface of dicotyledon leaves. Am. J. Botany, 23:255, 1936.
16. Stalfelt, M.G., Morphologic und anatomic des blattes als transpiration organ. Handbuch der Pflanzen Physiologie in Encyclopedia of Plant Physiology. Springer-Verlag, Berlin, Vol. 33, p. 324. Cited in Stewart, F.G., p. 629, 1959.
17. Parker, J., Effect of variations in the root-leaf ratio on transpiration. Plant Phys., 24:739, 1949.
18. Ketellaper, H.J., Stomatal Physiology. Ann. Rev. Plant Phys., 14:249, 1963.
19. Meyer, B.S. and D.B. Anderson, Plant Physiology, 2nd Ed., p. 203, Van Norstad Co., New York, 1955.
20. Robbins, W.W., T.E. Weier, and C.R. Stocking, Plant Physiology, John Wiley & Sons, New York, p. 187.
21. Hill, J.B., L.O. Overholts, H.W. Pop, and A.R. Groves, Botany, A Textbook for Colleges, 3rd Ed., McGraw-Hill, New York, p. 64, 1960.
22. Gates, D.M. and J.R. Hanks, Plant Factors affecting evapotranspiration, Chapter 27, p. 506, in Irrigation of Agricultural Lands, 1967.
23. Oleary, J.W. and G.N. Knocht, The effect of relative humidity on growth, yield and water consumption of beans. J. Am. Soc. Hort. Sci., 96:263, 1971.
24. Gates, D.M., Leaf temperature and energy exchange. Arch. Meteorological Geophy., 12:321, 1963.
25. Balba, A.M. and M.F. Soliman, Real and potential transpiration under different saline conditions. Alex. J. Agric. Res. 26:247, 1978.
26. Penman, H.L., Natural evaporation from open water bare soil and grass. Proc. Royal Soc. A., 193:120–145, 1948.
27. Blaney, H.F.and W.D. Criddle, Determining water requirements in irrigated areas from climatological and irrigation data. USDA, Soil Cons. Ser. T., p. 96, pp. 48, 1950.
28. Hargreaves, G.H., Consumptive use derived from evaporation pen data. Am. Soc. Civ. Eng. J. Irrig. Drainage Div., pap. 5863, IR. 1:97–105, 1968.
29. Rijtema, P.E., Evaporation from bare soil. 3.2.3 Eight Int. Course on Land and Drainage, Wageningen, Netherlands, 1969.
30. Van Bavel, C.H.M., Potential evaporation, The combination concept and its experimental verification. Water Resources Research, 2:445, 1966, In Agron. J., 66:450, 1974.
31. Meigs, P., Classification and occurrence of Mediterranean type dry climates. In Land use in Semi Arid Mediterranean Climates, UNESCO. Geog. Union Symp. Greece, UNESCO Pub. 1962.
32. Coe, M., The Conservation and Management of Semiarid Rangelands and Their Animal Resources. Chapter 8 in Techniques For Desert Reclamation, Goudie Ed., John Wiley & Sons, New York, 1990.
33. Harris, D.R., Tropical savanna environments: definition, distribution, diversity and development. In Human Ecology in Savanna Environments, Ed. D.R. Harris, pp. 3–27, Academic Press, London, 1980. Cited in Coe,[3] Chapter 8 of Techniques for Desert Reclamation, 1990.
34. De Martonne, E., Nouvelle carte mondiale de l'indice d'aridite. Ann. Geog. Vol. 51, No. 288:242, p. 407, 1942.
35. Koppen, W., Das Geographische System der Klimate. Vol. 1, Part C, Berlin, Cited in Abu El-Einein,[11] Climatic Geography, p. 40, 1936.
36. Thornthwaite, An Approach toward a national classification of climates. Geog. Rev. Vol. 38:55, 1948.
37. Emberger, L., Afrique du Nord-ouest, 1955, Cited by Meigs,[2] 1962.

38. Budyko, M.I., Climate and life, D.H. Millet (ed.), Academic Press, New York, p. 508, 1974.
39. Gresswell, K.R., Physical Geography, 1972. Longman, p. 58, Cited by Abu El-Einein, p. 387, 1981.
40. UNEP/FAO/UNESCO/WHO, Desertification Map of the World, A - CONF. 72-2, 1974.
41. MAB, Trends in research and in the application of science and technology for arid zone development. Man and Biosphere Tech. Notes No. 10, UNESCO, Paris, p. 53, 1979.
42. Kassas, M., Arid and semiarid lands. Problems and prospects, Agro-Ecosystems, 3:185–204, Elsevier, New York, 1977.
43. Ayyad, M.N. and S. Kamal, Distribution of plant species and growth forms in Western Mediterranean Desert of Egypt. Sahara Rev. 1:1–30, 1989.
44. Migahid, A.M., Water economy of desert plants. Bul. de l'Institute du Desert D'Egypt. Vol. 4, No. 1, 1, 1954.
45. Gul, A. and R.E. Allan, Interrelationships of seedling vigor criteria of wheat under different field situations and soil water potentials. Crop Sci. 16:515. Cited in Barnes, D.X. Managing root systems for efficient water use, Chapter 3 C of Limitations to Efficient Water Use in Crop Production, USA Inc., 1983.
46. Wright, I.N. and A.K. Dobrenz, Efficiency of water use and associated characteristics of Lehman lovegrass. J. Range Management, 26:210, 1972.
47. Miskin, K.E. and D.C. Rasmusson, Frequency and distribution of stomata in barley. Crop Sci., 10:575, 1970.
48. Chapman, V.J., Salt marches and salt deserts of the world. Interscience, New York, 1960. Cited in Ayyad and Kamal,[16] p. 8, 1990.
49. Walter, H., The adaptation of plants to saline soils. Tehran Symp., UNESCO Pub., Paris, p. 129, 1961.
50. Repp, G., The salt tolerance of plants. Basic research and tests. Tehran Symp., UNESCO Pub., Paris, p. 153, 1961.
51. Begg, J.E. and N.C. Turner, Crop Water Deficit. Adv. Agron. 161–217. Am. Soc. Agron., 1976.
52. Turk, K.J. and E.A. Hall, Drought adaptation of cowpea. II-Influence of drought on plant water status and relations with seeds yield. Agron. J. 72:421, 1980.
53. Radin, J.W., Physiological consequences of cellular water deficits. Osmotic Adjustment, Chapter 6 B of Limitation to Efficient Water Use In Crop Production. Am. Soc. Agron., 1983.
54. Soil Survey Staff, Keys To Soil Taxonomy, AID, USDA, Soil Management Support Services, pp. 279, 1987.
55. Kampen, J. and J. Burford, Production Systems. Soil Related Constraints and Potentials in the Semiarid Tropics with Special Reference to India. In Priorities for Alleviating Soil-Related Constraints to Food Production in the Tropics. Int. Rice Res. Inst. and N.Y. State Col. of Agric. and Life Science, Cornell University, p. 141, 1980.
56. Schantz, H.L., History and problems of arid land development, 1958. Cited in Kassas, M.,[15] 1977.
57. UNCOD Secretariat, Desertification: Its Causes and Consequences. Chapter 1. Pergamon Press, 1977.
58. Goudie, A.S. (Ed.), Techniques for Desert Reclamation, Chapter 1, by Goudie, John Wiley & Sons, New York, 1990.
59. Balba, A.M.

CHAPTER 2

Salt-Affected Soils

INTRODUCTION

The salt-affected soils are a major constraint for agricultural production in the arid regions. If their salt concentration allows economical plants to grow, their yields are low, cost of production is relatively high, and their need for water is substantially more than the nonsaline soils.

Investigations on a sound scientific basis had started since the 1930s to better understand the properties of these soils, their reclamation and suitable management. In the period 1930s to 1970s, a considerable number of soil research institutes were engaged in soil salinity and sodicity studies. The U.S. Salinity Laboratory was established and followed by several other specialized laboratories in various countries.

Several national, regional, and international meetings were held to exchange information and experience concerning the soil salinity and sodicity problems. The flow of published information reached its peak in the late 1970s. It is noticed that attention directed toward the salt-affected soils had tended to decrease since early in the 1980s. One reason for this is the attempt to use genetic engineering techniques to produce economical crops having the ability to tolerate soil and water salinity. Other reasons may be the accumulation of knowledge on this field and the need to concentrate research on other pressing problems, especially pollution of soils and water.

Primary and Secondary Salinity

Since ancient times man had noticed that there are soils that contain excessive amounts of salts. Plant growth on these soils is sporadic and production is lower than on soils without high salt content. Ultimately, land productivity diminishes, and farmers usually abandon the deteriorated land. The region becomes barren and eventually is added to the desert.

The sources of salt in the soil might be (Kovda[1]):

- Continental, due to weathering of igneous, or secondary rocks rich in salts.
- Marine sources where salts accumulate from the sea water in sea shores, especially in arid regions.
- Deltaic sources characterized with salt supply from the continent by means of rivers and from the sea in various periods.
- Anthropogenic sources which result from man's activities. Salinity due to the first three sources is usually termed Primary, while that due to the latter source is described as Secondary Salinity.

The primary salinity usually spreads out in poorly drained low-lying land in arid and semiarid regions. These lands receive considerable amounts of salts leached from elevated areas and accumulated in the slowly flowing groundwater basin sinks. If the water table is at or near to the soil surface, salts ascend by capillarity and evapotranspiration. A close relationship between the depth of groundwater, its salinity, and kind of salt accumulated in the soil exists (Kovda[1]).

Primary sodic soils receive salts from adjacent elevated areas and capillary flow of soil water. They are periodically water-logged and leached by rain which reduces their concentration of soluble salts (Rhoades[2]).

Sodic soils develop when the soil is subjected for a long time to water having SAR > 5 and/or light proportion of HCO_3 salts. As the water is exposed to evapotranspiration, the SAR increases by a factor equal to the square root of the volume decrease factor. The removal of Ca from solution during evaporation—as by precipitation—will further increase the SAR value (Bower[3]).

In arid and semiarid conditions "P/ET" (the ratio of precipitation to evapotranspiration) is 0.03 < P/ET < 0.2 and 0.2 < P/ET < 0.5, respectively, as stated in Chapter 1. Introduction of irrigation causes several ecological, chemical, and hydrological changes. Instead of the low rainfall and scarce vegetation, ample amounts of water flow through the soil, and green cover resulting from intensive agricultural activities prevails the year round. These new conditions change the natural water balance in the newly irrigated areas causing the groundwater level to rise up with waterlogging, salinization, or sodification unless drainage is efficient.

More recently, it was noticed that clearing of lands for dryland agriculture has become a recognized source of soil salinization (Rhoades[2]).

From the above-stated studies[1-3] and others (Doering and Sandoval[4]; Balba and Soliman[5,7]; Ayers and Westcot[6]; Eaton[8]) waterlogging, salinization, and sodification are the results of the following conditions which cause waterlogging:

- Flooding by seasonal river floods or rains.
- Seepage of water from irrigation canals.
- Rise of groundwater table due to excessive irrigation.
- Presence of impermeable layers in the soil profile.
- Low-lying areas surrounded by or adjacent to areas of relatively elevated levels usually receive seepage water from those surrounding areas, Doering and Sandoval.[4]

When upward movement of the groundwater table reaches the soil surface by capillarity and evaporates, salts are left behind on the ground surface.

Lack of fresh water is an obstacle confronting development in many countries of the arid and semiarid regions. It has always been thought in these regions that drainage water be reused for irrigation to partially satisfy the need for water.

In arid and semiarid irrigated regions, the following conditions enhance the salinization processes:

- Shallow groundwater table.
- High salt concentration in the groundwater.
- Medium soil texture because capillarity in this soil texture is faster than in other textures, Balba and Soliman.[5]
- Long periods between each irrigation.
- Subsurface irrigation, Ayers and Westcot[6]; Balba and Soliman.[7]
- Inefficient drainage system.

- Water regime of crops (cereals vs. rice).
- Quality of irrigation water.
- Fallowing in dry hot seasons.

The following conditions enhance soil sodification:

- Leaching the saline sodic soils without gypsum application if the soil content of Ca is not satisfactory.
- Use of irrigation water having SAR value of 12 or 8 in case of montmorillonite, illite-vermiculite- and sesquixides-rich soils, respectively. Also, waters rich in HCO^{3-} and/or CO_3^- precipitate Ca and increase sodification hazards, Eaton.[9]

Formation of Sodium Carbonate in Soils

Groundwater containing soluble carbonate and bicarbonate is an important factor in the formation of sodic soils in various regions (El-Gabaly[9]; Whitting and Janitzky[10]; Beek and Breemen[11]).

Sodium carbonate may be formed in the soil system as a result of several processes among which are the following:

Weathering of Aluminum Silicates

Some of the silicate minerals are easily weathered in the presence of water and, especially, of carbonic acid. The chemical weathering of these minerals results in sodium and potassium carbonates and bicarbonates.

Decomposition of Plant Residues

The microbial activities

The denitrification and desulfurication processes take place under anaerobic conditions. The following factors enhance the desulfurication process:

- Anaerobic condition.
- Presence of soluble sulfate.
- Presence of organic matter as an energy source.
- Presence of the microbe desulfovibrio which has the ability to reduce the sulfate to use its oxygen.

According to Whitting and Janitzky,[10] desulfurication in the presence of organic matter results in the formation of sodium carbonate. Thus, divalent cations' proportion in the soil solution is decreased.

Other processes also form sodium carbonate as suggested early by Kelley[12] when NaCl or Na_2SO_4 solutions rise to the soil surface—with capillarity—the Na-clay proportion increases. When the soil is leached with river floods or rains, the Na-clay hydrolyzes giving Na_2CO_3.

Kovda[13] summarizes the forms of anthropogenic (secondary) salinization as follows:

In Non-Irrigated Land

1. Formation of secondary saline soils as a result of overgrazing and compactness of sod land:

   a) along the contact belt of mountainous foothills and plains.
   b) on low terraces of a valley, after river flooding was ended by dam construction.

2. Formation of secondary saline soils as a result of the disposal of brackish water pumped from:

   a) petroleum wells.
   b) coal mines.
   c) industrial plants.

3. Salinization of soils after sea water invasion under the influence of land subsidence or after the destruction of a dike protecting (and against the sea) or after heavy storms or earthquakes. (The anthropogenic role in this mechanism is not clear.)
4. Accelerated formation of saline alluvial soils on deltaic and tidal wave areas after periodic floodings have stopped because of dam construction in the middle and upper reaches of a river valley.

In Irrigated Land

1. Formation of water-logged and saline soils, along the network of non-lined canals as a result of water seepage, groundwater elevation and subsequent evaporation.
2. Formation of spotted saline fields, then totally saline fields, after several (5-10-15) years of irrigation of land without appropriate drainage installations for the removal of saline subsoil waters.
3. Wrong application for watering soils using brackish or alkaline irrigation water taken from:

   a) saline water rivers.
   b) tube wells installed into saline groundwater or after the overpumping of good subsoil water.
   c) sea or gulf sources. (This mechanism does not exist yet.)

THE SIZE OF THE PROBLEM

Efforts to appraise the problem of salinization, sodification, and waterlogging in the world are based on estimation of the areas suffering from and subject to deterioration by these processes from various sources.

Dregne[14] estimated the global area affected by waterlogging and salinization to be 21 and 20 M ha, respectively.

From the Desertification Map of the World (FAO/UNESCO[15]) the surface area in the six continents which is considered subject to salinization and sodification was estimated by the writer (Balba[16]) and found to be about 6 M km^2 or about 600 M ha. They are divided among the continents as follows (in km^2):

Table 2.   Chemical characteristics of saline, nonsaline sodic, and saline sodic soils.

Soil	EC dS m⁻¹	ESP	pH
Saline	> 4.0	< 15	< 8.5
Sodic (nonsaline)	< 4.0	> 15	> 8.5
Saline sodic	> 4.0	> 15	< 8.5

Source: Richards.[26]

No. America, 260,000
So. America, 600,000
Africa, 300,000
Asia, 3,400,000
Australia, 1,400,000
Europe, 14,000

This area was classified into 3 degrees of desertification hazards: very high, high, and moderate. Only about 20,500,000 ha are considered subject to a very high degree of desertification hazards due to salinization and sodification.

In the FAO[17] report "Water for Agriculture" the irrigated areas were considered prone to salinization, sodification and/or waterlogging unless they were equipped with drainage.

The following values were reported, concerning Africa, Latin America, Near East, and Asia:

- Actual (1975) equipped irrigated areas: 91,986,000 ha
- Projected (1990) equipped irrigated areas: 114,190,000 ha
- Irrigation improvement targets, 1975–1990:
  minor improvement: 26,853,000 ha
  major improvement: 18,135,000 ha
  Total: 44,988,000 ha
- Area equipped with drainage (1975): 134,432,000 ha
- Drainage improvement on:
  irrigated land: 52,423,000 ha
  non-irrigated land: 25,761,000 ha
  Total: 78,184,000 ha

Massoud[18] compiled data from the FAO/UNESCO Soil Map of the World concerning the areas suffering from salinity, sodicity and/or waterlogging in the developing countries (Table 2). He stated that information on the exact areas degraded or their degree of deterioration is not available for all the countries affected. He reported the figures shown in Table 3.

From published information, the salt-affected soils are important constraints for agricultural production in the following countries in almost all continents: In North America, the soil salinity is spread out in 17 western states of the United States. Also, soils in the west of Canada are similar to those in western United States. Ehlrich and Smith[19] described the solonetz soils in Manitoba. In Cuba, Blazhni[20] classified the soil in Cuba Delta to marine saline soils, meadow swamps, and meadow solonchak depending on the vicinity to the sea and the altitude. The South America coast in Peru 2000 km long and 10–15 km wide is mostly saline. Salts of marine origin are spread out in Lablazos Marinis in the north and the middle and the Pampas Region. Also secondary salinization spreads out in the cultivated areas in Peru year after year due to the irrigation system and lack of

Table 3. Calcium concentration (Ca_x) expected to remain in near-surface soil-water following irrigation with water of given HCO_3/Ca ratio and EC_w.[1,2]

Ratio of HCO₃/Ca	Salinity of Applied Water (EC_w) (dS/m)											
	0.1	0.2	0.3	0.5	0.7	1.0	1.5	2.0	3.0	4.0	6.0	8.0
.05	13.20	13.61	13.92	14.40	14.79	15.26	15.91	16.43	17.28	17.97	19.07	19.94
.10	8.31	8.57	8.77	9.07	9.31	9.62	10.02	10.35	10.89	11.32	12.01	12.56
.15	6.34	6.54	6.69	6.92	7.11	7.34	7.65	7.90	8.31	8.64	9.17	9.58
.20	5.24	5.40	5.52	5.71	5.87	6.06	6.31	6.52	6.86	7.13	7.57	7.91
.25	4.51	4.65	4.76	4.92	5.06	5.22	5.44	5.62	5.91	6.15	6.52	6.82
.30	4.00	4.12	4.21	4.36	4.48	4.62	4.82	4.98	5.24	5.44	5.77	6.04
.35	3.61	3.72	3.80	3.94	4.04	4.17	4.35	4.49	4.72	4.91	5.21	5.45
.40	3.30	3.40	3.48	3.60	3.70	3.82	3.98	4.11	4.32	4.49	4.77	4.98
.45	3.05	3.14	3.22	3.33	3.42	3.53	3.68	3.80	4.00	4.15	4.41	4.61
.50	2.84	2.93	3.00	3.10	3.19	3.29	3.43	3.54	3.72	3.87	4.11	4.30
.75	2.17	2.24	2.29	2.37	2.43	2.51	2.62	2.70	2.84	2.95	3.14	3.28
1.00	1.79	1.85	1.89	1.96	2.01	2.09	2.16	2.23	2.35	2.44	2.59	2.71
1.25	1.54	1.59	1.63	1.68	1.73	1.78	1.86	1.92	2.02	2.10	2.23	2.33
1.50	1.37	1.41	1.44	1.49	1.53	1.58	1.65	1.70	1.79	1.86	1.97	2.07
1.75	1.23	1.27	1.30	1.35	1.38	1.43	1.49	1.54	1.62	1.68	1.78	1.86
2.00	1.13	1.16	1.19	1.23	1.26	1.31	1.36	1.40	1.48	1.54	1.63	1.70
2.25	1.04	1.08	1.10	1.14	1.17	1.21	1.26	1.30	1.37	1.42	1.51	1.58
2.50	0.97	1.00	1.02	1.06	1.09	1.12	1.17	1.21	1.27	1.32	1.40	1.47
3.00	0.85	0.89	0.91	0.94	0.96	1.00	1.04	1.07	1.13	1.17	1.24	1.30
3.50	0.78	0.80	0.82	0.85	0.87	0.90	0.94	0.97	1.02	1.06	1.12	1.17
4.00	0.71	0.73	0.75	0.78	0.80	0.82	0.86	0.88	0.93	0.97	1.03	1.07
4.50	0.66	0.68	0.69	0.72	0.74	0.76	0.79	0.82	0.86	0.90	0.95	0.99
5.00	0.61	0.63	0.65	0.67	0.69	0.71	0.74	0.76	0.80	0.83	0.88	0.93
7.00	0.49	0.50	0.52	0.53	0.55	0.57	0.59	0.61	0.64	0.67	0.71	0.74
10.00	0.39	0.40	0.41	0.42	0.43	0.45	0.47	0.48	0.51	0.53	0.56	0.58
20.00	0.24	0.25	0.26	0.26	0.27	0.28	0.29	0.30	0.32	0.33	0.35	0.37
30.00	0.18	0.19	0.20	0.20	0.21	0.21	0.22	0.23	0.24	0.25	0.27	0.28

1. Assumes Ca as CaCO₃ or silicates, no precipitation of Mg and partial pressure of CO_2, near the soil surface (P_{co2}) is 0.0007 atm.
2. Ca_x, HCO_3, Ca are reported in meq/l; EC_w is in dS/m. Source: Suarez.[38]

drainage which causes waterlogging. With warm dry summer, salts accumulate on the soil surface. According to Zavaleta,[21] production of about 25-30% of the cultivated area was decreased due to salt accumulation. In Asia, Kovda[1] stated that the salt-affected soils are spread out in China in the Sungari River Valley and Delta of Hwan Ho River, especially near the Yellow Sea. The salt-affected soils in Azerbaijan are a result of weathering processes as stated by Agaev.[22] In Kazakhstan, Borovski[23] stated that the soil is rich in Na_2CO_3 which is formed from the decomposition of plant material, while in Tienchan Valley it is a result of geochemical transformation in the deep groundwater. In River Ili Valley, sodium carbonate was formed as a result of granite weathering in the Middle Mountains. According to Kovda,[1] saline soils cover vast areas of the river valleys in east and west of Siberia, the Oral Region and Arexes valley in Arminea. Also, these soils are present in Kukasia, Okrania, and Muldavia in the deposits of Rivers Don, Deneper, and Danub. The salt-affected soils are also found in East Europe, especially Hungary where they cover about 10% of the country area, Czechoslovakia, Yugoslavia (325,000 ha), and Rumania (300,000 ha) (Szabolcs[24]). Salt-affected soils cover about 600,000 ha in Aragon, Catalonia and Andalucia in Spain (Ayers[25]). In the Netherlands, most of the land was submerged with sea water and was dried and leached.

According to Rhoades,[2] about 10% of the total land area of the world suffers from salt accumulation. According to the estimates of FAO/UNESCO,[15] as much as half of the area of all existing irrigation systems of the world are seriously affected by salinity and/or waterlogging. The economic and social repercussions of soil salination are felt most acutely by the population of arid regions which depend primarily upon irrigated agriculture.

CHARACTERISTICS OF SALT-AFFECTED SOILS

Morphological Characteristics

Man distinguished the salt-affected soils from their appearance before defining their chemical and physical characteristics.

He noticed that such unproductive soils are usually covered during dry periods with an afforescence of salt crust. They may be wet, fluffy, solid, and of light or dark color. Later, he could know that differences in the crust color and nature are related to the kind of salt prevailing in the soil. Thus he knew that $MgCl_2$ salt gives rise to a dark crust highly hygroscopic and absorbs atmospheric moisture. A mixture of sodium chloride and sulfate results in a crystalline white mass. Powdery fine fluffy, loose, and dusty surface characterizes the abundance of $CaCO_3$ and $CaSO_4$. The dark-colored surface indicates the presence of alkali soils, termed in some regions as Soda or Sodasolonchak soils. The dark color of the "black-alkali" soils was related to the presence of dilutions of humus and organic colloids. The profiles which show a dark bluish color might indicate the presence of a shallow water table.

Man also noticed that these soils are barren or, when salt concentration allows plant growth, the growing plants are sporadic, unhealthy, and usually stunted and may have a deep bluish green color. The yields are usually low unless the crop is salt-tolerant. The natural vegetative cover is usually made up of halophytic plants.

Chemical and Physical Characteristics

Measuring Soil Salinity

The electrical conductivity (EC) of the soil water extract at saturation is directly related to the EC of the soil solution. Saturation percentage is about twice the soil moisture at maximum water-holding capacity and about four times that at the permanent wilting point PWP (moisture content at 15 atmosphere). Thus, EC—or the soluble salt concentration—of the saturation extract would be about one half of the soil water at maximum water-holding capacity and one fourth of EC of soluble water at wilting point (Richards[26]).

Determination of EC is performed by placing the soil extract between two electrodes having constant geometry and constant distance of separation. The resistance to the electric current passing between the two electrodes is inversely proportional to the solution salt concentration. Thus, the conductivity of the current passing through the solution is the reciprocal value of its resistance. The resistance unit is the Ohm, thus the conductivity unit was termed Moh. At present, Siemens is the standard unit of conductance and when expressed per unit of distance between the two electrodes the standard unit of conductivity is Siemen per meter. Since this unit is too large for most saturation paste extracts conductivities, deci Siemen (dS per meter at 25°C) is commonly used.

An approximate numerical relationship between EC in dS m^{-1} and the solution concentration was found as follows:

$$\text{Solution concentration, meq/l} = EC \times 10 \text{ (U.S. Sal. Lab.)}[26]$$

Also, the osmotic pressure (OP) of the solution was found to be approximately OP = 0.36 $EC_e \times 10^3$ (U.S. Sal. Lab.[26]).

The Four-Electrode Probe

Rhoades and Ingvalson[27] developed the four-electrode probe for measuring bulk soil electrical conductivity in the field. Thus, there will be no need for extensive soil sampling and laboratory manipulation.

The equipment is made up of a source for electrical current, a conductivity meter, stainless steel or brass electrodes, wires for connections, a meter for measurement, and a thermometer. The equipment measures the electrical resistance in the bulk soil after adjusting to 25°C, the soil moisture and soil texture.

The readings of electrical resistance are transformed to EC values for the soil solution. Rhoades[28] has prepared this transformation.

This equipment can be of significant help in recording and monitoring variations in soil salinity. Leaching and drainage operations can also be followed up.

Rhoades et al.[29] developed a method to determine EC_e (electrical conductivity of the soil saturation extract) from the electrical conductivity of the saturated soil paste, EC_p. They determined the saturation percentage of mineral soils in the field for purposes of salinity appraisal from the weight of the paste-filled cup and developed (Figure 1) for this purpose.

They derived the following equation:

$$EC_p = \left[\frac{(Q_s + Q_{ws})^2 \, EC_e \, EC_s}{(Q_s) \, EC_e + (Q_{ws}) \, EC_s} \right] + (Q_w - Q_{ws}) \, EC_e \qquad (10)$$

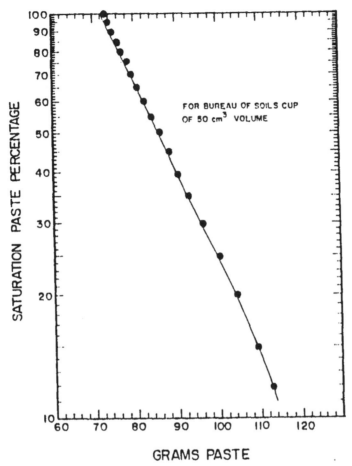

For Bureau of Soils Cup of 50 cm³ Volume

Figure 1. Theoretical relationship between saturation percentage (SP) and weight (in grams) of 50 cm³ of saturated soil paste, assuming a particle density of 2.65 g/cm³ (after Rhoades).

where EC_p and EC_e are electrical conductivity of soil paste and of the paste extract respectively, Q_w and Q_s are the volume fraction of total water and solids in the paste, respectively, Q_{ws} is the volume fraction of water in the paste that is coupled with the solid phase to provide a series-coupled electrical pathway through the paste, EC_s is the average specific electrical conductivity of the solid particles, and the difference $(Q_w - Q_{ws})$ is Q_{wc}, which is the volume fraction of water in the paste that provides a continuous pathway for electrical current to flow through the paste (a parallel pathway to Q_{ws}).

Assuming the average particle density (e_s) of mineral soils to be 2.65 g/cm³ and the density of saturated soil paste extracts (e_w) to be 1.00, Q_s and Q_w are directly related to SP as follows:

$$Q_w = SP \ / \ [(100/e_w \ e_s) + SP] \tag{11}$$

$$Q_s = 1 - Q_w \tag{12}$$

As shown by Rhoades et al.,[29] saturation percentage of mineral soils, generally, can be adequately estimated in the field for purposes of salinity appraisal. The weight of the paste-filled cup (Figure 1) may be used for this purpose.

EC_e can be determined from measurement of EC_p and SP (using Equation 10) if values of e_s, Q_{ws}, and EC_s are known. These parameters can adequately be estimated, as demonstrated by Rhoades et al.,[29] for typical arid land soils of the southwestern United States e_s may be assumed to be 2.65 g/cm^3. EC_s may be estimated from SP as: $EC_s = 0.019(SP) - 0.434$. The difference $(Q_w - Q_{ws})$ may be estimated from SP as:

$$(Q_w - Q_{ws}) = 0.0237 \ (SP)^{0.6657} \tag{13}$$

Determination of the soil saturation paste conductivity using the EC cup (Rhoades et al.[29]).

Apparatus

1. Portable balance, capable of weighing accurately to the nearest one gram.
2. Conductivity meter, temperature-compensating type.
3. Cup-type conductivity cell, 50 cm^3 volume such as the Bureau of Soils cup (U.S. Sal. Lab.[26]).

Reagents

Standard KCl solutions, 0.010 and 0.100 N solutions:

$$0.010 \ \underline{N} \text{ has EC} = 1.412 \text{ dS m}^{-1} \text{ at } 25°C$$
$$0.100 \ \underline{N} \text{ has EC} = 12.900 \text{ dS m}^{-1} \text{ at } 25°C$$

Procedure

Rinse and fill the conductivity cup with KCl solution. Adjust the conductivity meter to read the standard conductivity. Rinse and fill the cup with the saturated soil paste, tap the cup to dislodge any air entrapped within the paste. Level off the paste with the surface of the cup. Weigh the cup plus paste; subtract the cup tare weight to determine the grams of part occupying the cup, obtain the SP value from Figure 3, paper 4,4 (Rhoades et al.[29]) corresponding to this weight. Connect the cup electrodes to the conductivity meter and determine the EC, corrected to 25°C directly the meter display. Obtain EC_s from Figure 2 from EC_p using the curve corresponding to the SP value or as calculated from Equations 10–12.

Mobile Soil Salinity Assessment

Spatial variability and dynamic nature of the soil salinity, caused by the effects and interactions of various factors stated above, complicate its mapping and monitoring. Rhoades[30] developed rapid instrumental techniques for measuring soil electrical conductivity (EC_a) as a function of spatial position on the landscape coupled with procedures for inferring salinity from EC_a with computer-assisted mapping techniques capable of associat-

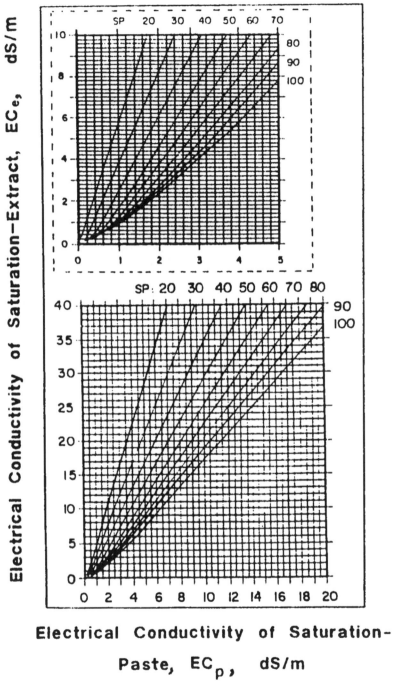

Figure 2.   Electrical conductivity of saturation paste, EC_p, dS/m (Rhoades et al.).

ing and analyzing spatial databases, and with appropriate spatial statistics to create an integrated system which has the potential of meeting salinity assessment needs.

The equipment and methodology of the first generation mobile automated instrumental systems developed in 1992 were modified in 1993 to increase depth and volume of soil that

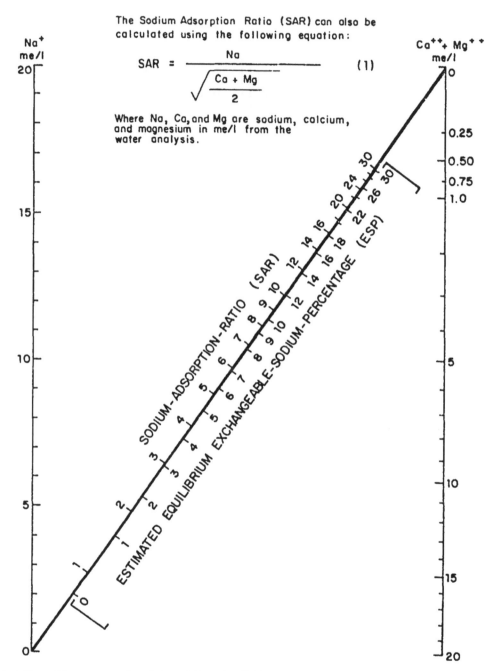

The Sodium Adsorption Ratio (SAR) can also be calculated using the following equation:

$$SAR = \frac{Na}{\sqrt{\dfrac{Ca + Mg}{2}}} \qquad (1)$$

Where Na, Ca, and Mg are sodium, calcium, and magnesium in me/l from the water analysis.

Figure 3.   Nomogram for determining the SAR value of irrigation water and for estimating the corresponding ESP value of a soil that is at equilibrium with the water (Richards 1954).

can be measured. A description of the equipment and procedure of its utilization were presented in Rhoades' report.[157]

A soil is considered "salt-affected" when its paste extract has electrical conductivity $EC = 4.0$ dS m^{-1} or more at 25°C. Generally, the salt-affected soils have a salt crust or a dark powdery cover as described above.

The chemical and physical characteristics of salt-affected soils are determined by the kinds and amounts of salts present. In case of a soil classified as "saline," soluble and exchangeable potassium are generally minor constituents. Sodium seldom comprises more than half of the soluble cations and accordingly does not constitute a significant proportion of the cation exchange capacity. The relative amounts of calcium and magnesium present in the soil solution and on the exchange complex may vary considerably. When the saline soil content of salts is dominated by sodium—and not by calcium plus magnesium—the exchangeable sodium constitutes a major proportion of the soil cation exchange capacity (generally $> 15\%$) and the soil is classified as saline sodic or alkali. Because both "saline" and "saline sodic" soils contain excess salts, the soil particles are flocculated and a soil permeability problem does not exist (U.S. Sal. Lab.[26]).

According to the U.S. Sal. Lab.,[26] the soil is classified "nonsaline sodic" when it contains exchangeable Na more than 15% of its cation exchange capacity accompanied with low soluble salt content (EC < 4.0 dS m^{-1}). The soil pH under these conditions may be above 8.5.

Nonsaline sodic soils may develop by time a dense impermeable clay layer of dispersed Na clay underneath a relatively coarse friable soil surface (solonetz soil). If sufficient time has not elapsed for such a developed profile, the presence of excess exchangeable sodium may have a marked influence on the soil's physical properties. As its proportion increases, the soil tends to become more dispersed and the permeability problem is accentuated.

The Permeability Problem of the Nonsaline Sodic Soils

Leaching the saline sodic soil poor in Ca, the soluble salts decrease and the soil may become "nonsaline sodic" with a high content of exchangeable Na and a pH above 8.5. The soil particles are dispersed and the soil becomes unfavorable for the entry and movement of water and for tillage when dry.

In saline soils, the sulphate and chloride anions prevail whereas in the nonsaline sodic soils, bicarbonate and carbonate anions prevail. In the solonetz soils, the carbonates are frequently missing.

The U.S. Sal. Lab.[26] classifies the soils according to their soluble salt content expressed in electrical conductivity units of their soil paste extract and exchangeable Na percentage (ESP) as shown in Table 2 (Richards[26]).

The adsorbed cations in the soil system are attracted by the clay negative charge to the clay surfaces and at the same time tend to diffuse into the solution away from the clay surfaces under concentration gradient. As a result of the two opposing forces, the adsorbed cations concentration decreases with distance from the clay surfaces out into the bulk solution. The force that attracts divalent cations to the clay surfaces are twice that attracting the monovalent cations. Accordingly, the divalent cations surrounding the clay are more compressed toward the clay than the monovalent.

The clay particles form aggregates under van der Waals forces. When the cations surrounding the clay particles are compressed, the van der Waals forces between clay particles are enhanced. On the other hand, these forces are diminished when the surrounding cations are loose. The clay particles surrounded by divalent cations can approach sufficiently close resulting in a more porous aggregate structure. Repulsion between clay particles allows more solution to fill in between them causing swelling which reduces the size of the interaggregate pore spaces and less permeable soil. Montmorillonitic soils having ESP values more than about 15% are more subject to swelling. These minerals preferentially adsorb exchangeable Na on their external surfaces which make up about

15% of their cation exchange capacity (EC). With further increases in adsorbed Na, it enters the interlayer positions between the parallel platelets where it creates repulsion forces causing swelling. Dispersion and slaking may occur at ESP values less than 15% when the electrolyte concentration is sufficiently low. Soil solution composed of high solute concentrations and calcium and magnesium salts produce good soil physical properties.

Low concentrations and sodium salts adversely affect permeability and tilth (Rhoades[2]). Abrol et al.[31] stated that a survey of published data showed that for sodic soils, most often an ESP of 15 to 20 is associated with saturation paste pH of 8.2. Therefore, for diagnostic purposes they suggested that a saturation paste pH of 8.2 will be more realistic than the value of 8.5 which is nearly always associated with a higher value of exchangeable sodium ratio (ESR). The clay content of saline soils is usually high, although they may be found in various textural classes. Because of the presence of excess salts, the soil particles in the saline or saline sodic soils are flocculated and the soil is permeable. The nonsaline sodic soils have deflocculated particles of water unstable structure. They form hard and compact clods when dry. Thus they are difficult to till when dry. They are elastic and sticky and take a long period to dry.

The solonetz soils stated above are characterized by a structural clay horizon of well developed columns.

To chemically diagnose the saline soils, the U.S. Sal. Lab. (Richards[26]) used to consider the soil saline, when its paste extract has EC 4 dS m^{-1} or more at 25°C. The concentration of paste extract was used because of the above stated relationship between the paste content of water and the soil water content at the field capacity level. Thus, it is understood that the salt concentration in the soil water at the field capacity is about double that determined in the paste extract. The level of EC 4 dS m^{-1} was chosen to characterize the saline soils, from the findings of the U.S. Sal. Lab.[26] that the sensitive plants which grow in soils having paste extracts with EC 4 dS m^{-1} will lose half of their yield due to the salt content. In a more detailed classification, the soils were classified according to the harmful effect of the salinity level of their paste extract on different plants into 5 classes as follows:

Not harmful   EC 0 – 2 dS m^{-1}
Harmful to sensitive plants   EC 2 – 4 dS m^{-1}
Harmful to most plants   EC 4 – 8 dS m^{-1}
Only tolerant plants can grow   EC 8 – 16 dS m^{-1}
Only very tolerant plants can grow   EC > 16 dS m^{-1}

Evidently, these limits are set to facilitate the evaluation of the soils from the salinity standpoint.

The Soluble-Exchangeable Cations Relationships

Attempts have been made to establish mathematical relationships between soluble and exchangeable cations in soil systems. These attempts are usually based on different theoretical concepts expressing the cation exchange reaction, though the Gapon isotherm is most commonly used. Difficulties have always been encountered in this field because (1) most of the suggested equations were established for systems containing two cations while in the soil systems several cations are present, (2) the soils are mixtures of different clay mineral types, (3) the determination of exchangeable Ca and Mg in $CaCO_3$-rich soils is not satisfactorily accurate, and (4) the cation activities which control the exchange reaction differ in soil systems according to the anion composition of the soil solution (U.S. Sal. Lab.[26]; Oster and Sposito[32]; Rao et al.[33]; Balba and Balba[34,35]).

Several investigators attempted to improve the mathematical expression adopted by the U.S. Sal. Lab.[26] by using cation activities instead of total cation concentrations.

Sposito and Mattigod[36] used SAR^x, the sodium adsorption ratio based on molar free-ionic concentrations and E_{Na}, the equivalent fraction of Na^+ in the exchange phase (the ratio of the charge of exchangeable Na to the total charge expressed by the cation exchange capacity), i.e., $100 \, E_{Na}$. For the case of Na^+-Ca^{2+} exchange, the correct expression according to Sposito and Mattigod[36] is:

$$K_G = (SAR^x \, / \, 2E_{Na}) \, (1 - E^{2_{Na}})^{1/2} \tag{14}$$

Assuming $(1 - E^2_{Na})$ could be replaced by $(1 - E_{Na})$, Sposito and Mattigod[36] derived the following relationship:

$$ESP = \frac{(100 \, / \, 2K_G) \, (SAR^x)}{1 + (SAR^x \, / \, 2K_G)} \tag{15}$$

Equation 13 is of the same form as the ESP-SAR relation introduced by the U.S. Sal. Lab.[26] in 1954, though it contains SAR^x instead of $SAR = Na_T / [Ca_T]^{1/2}$ which is based on total molar concentrations and KG instead of the traditional Gapon selectivity coefficient.

Oster and Sposito[32] ended their attempts by stating that "the widely-used empirical relationship between ESR (or ESP) and SAR developed in 1954 by the U.S. Sal. Lab. Staff is indistinguishable from the ESR-SAR relation predicted by an approximate calibrated form of eq. 14 over the SAR range of practical significance." However, for a single soil, a relation between ESR and the activity ratio $(Na^+)/(Ca^{2+})$ is on firmer thermodynamic grounds than one between ESR and either SAR^X or SAR (Rao et al.[33]; Nakayama[37]). This is especially important when the solution Phase is rich in SO_4^- anions. Under this condition ESP predicted from equations by using total cation concentration would be lower than when activities are used (Rao et al.[33]; Balba and Balba[34]).

The value of Ca in the sodium adsorption ratio SAR is not constant. It is affected by CO_2 pressure and HCO_3^- concentration in the soil water. The U.S. Sal. Lab.[26] old classification of water in 1954 used to take into consideration the residual $(CO_3^- + HCO_3^-)$ which equals $[(CO_3^- + HCO_3^-) \, (Ca^{2+} + Mg^{2+})]$ as originally suggested by Eaton.[8] Suarez[38] introduced the adjusted sodium adsorption ratio:

$$Adj \, R_{Na} = \frac{Na}{(Ca_x + Mg)/2} \tag{16}$$

where Na is sodium in the soil water or in the irrigation water in meq/l, Ca_x is a modified calcium value obtained from Table 3, meq/l, and Mg is magnesium in the irrigation water or soil water.

The value of Ca_x is the expected Ca concentration in the irrigation water in the presence of various HCO_3/Ca ratios ($PCO_2 = 0.0007$ atm.) and EC values, as the salinity increases the solubility of the Ca minerals increases.

Ayers and Westcot[6] showed that R_{Na} is generally very close to the SAR for various types of water around the world when SAR is calculated after adjustment of Ca concentration using Equation 16.

Table 4.   Concentration factors (x) for predicting soil salinity (EC$_e$)[1] from irrigation water salinity (EC$_w$) and the leaching fraction (LF).

Leaching Fraction (LF)	Applied Water Needed (% of ET)	Concentration Factor[2] (x)
0.05	105.3	3.2
0.10	111.1	2.1
0.15	117.6	1.6
0.20	125.0	1.3
0.25	133.3	1.2
0.30	142.9	1.0
0.40	166.7	0.9
0.50	200.0	0.8
0.60	250.0	0.7
0.70	333.3	0.6
0.80	500.0	0.6

[1]   The equation predicting the soil salinity expected after several years of irrigation with water of salinity EC$_w$ is

$$EC_e \ (dS/m) = EC_w \ (dS/m) \bullet x$$

[2]   The concentration factor is found by using a crop water use pattern of 40-30-20-10.
Source: Ayers and Westcot.[6]

Prognosis of Salinity and Sodicity

Researchers have paid more attention to identifying the salt-affected soils and their characteristics than to predicting the possibilities of salinization or sodification of a particular area or project. This might be due to:

- The importance of diagnosing the soil in the stage of selecting the project site.
- The methods and steps of land reclamation differ according to the soil diagnosis so that these methods should suit the soil characteristics.
- The role of salinity or sodicity prognosis in applying the necessary measures to protect the land was not well understood.
- The complexity of the factors that may cause soil deterioration and the difficulties to quantitatively evaluate the collective effect of these factors.
- Attention to the project is decreased compared with the attention given at its beginning.

Land reclamation projects have been carried out in several parts of the world. Many of these projects are located in arid regions and had to depend on irrigation. Incidents of land deterioration due to salinization have been replete which signifies losses of considerable investments, efforts, and hopes for more food and fiber. Deterioration is not usually limited to the reclaimed area but usually is extended to the other neighboring lands. Because of these losses, FAO[39] called for expert consultation to obtain advice on scientific and applied research activities in the prognosis of salinity and alkalinity with the objective of preparing a guideline for predicting salinity and alkalinity and to make recommendations concerning the measures needed for implementation of the prognosis concept.

In order to be able to predict the future of the land of the project and possibilities of its deterioration under the prevailing climatic conditions and the expected technologies to be applied, the following information and studies are needed:

- Environmental conditions which may lead to soil deterioration.
- The project soil characteristics determined in the planning stage.
- The information collected concerning the region and similar projects during an extended period of time.

The following sources of information may be consulted:

- The soil survey reports and maps.
- Hydrology and hydrogeology reports on the region.
- Field and laboratory studies.
- Aerial photographs and Landsat reports.
- Models to simulate the major processes.
- The climatic records have an important role. Without these records, the rest of the collected information would be valueless and may lead to erroneous interpretation.

The investigator should give the following points special consideration:

1. The main purpose of the study is prognosis and not more diagnosis of the land.
2. Some soil characteristics may change with irrigation and cropping and others may not.
3. Land deterioration may be a result of properties which were not covered during the diagnostic study, such as presence of impermeable layer so deep in the profile that regular soil survey does not take it into consideration, or a result of the neighboring land conditions.
4. Prognosis of soil salinity or sodicity requires various factors which influence each other. Thus, it is necessary to consider all information mentioned above and their collective effect on salinization or alkalization of the project soil.
5. Experts should be consulted.

STUDIES REQUIRED FOR PREDICTING SALINIZATION DUE TO UPWARD MOVEMENT OF GROUNDWATER AND ITS SUBSEQUENT EVAPORATION (Balba[40])

Groups of Studies

Groundwater Studies

- Depth from soil surface.
- Salt concentration and composition.
- Hydraulic properties including flow rate, gradient, and direction.

Soils Studies

- Morphological study of the soil profile.
- Texture and structure.

- Soil-water constants.
- Presence of impermeable layers, their depth from the soil surface and their chemical composition.
- The hydraulic conductivity.
- The exchangeable cations.
- Concentration and composition of the soluble salts in the soil paste extract.

Irrigation Studies

- Method of surface or subsurface irrigation.
- Design of the irrigation canals and drainage network.

Agronomic Studies

- The expected cropping system.
- Water consumption of each crop and crop rotation.
- Irrigation regime of each crop.

Hydrological Studies

- Seepage of water from neighboring areas.

From these studies, it should be possible to calculate the upward movement rate of the groundwater and calculate the time required to be effective in the deterioration of the project soil.

Predicting Soil Salinization due to Upward Movement of Groundwater and its Subsequent Evaporation

The above stated groups of investigations which constitute the basic elements of the problem of waterlogging are the same for predicting soil salinization.

Predicting Soil Salinity and Sodicity Resulting from Saline Water Utilization

Guidelines of saline water utilization for irrigation are stated elsewhere in this book. Applying these standards should indicate the possibilities of soil salinization as a result of using the water for irrigation. However, these standards deal only with the water characteristics. The effect of utilizing a particular water on a particular soil depends, in addition to the water properties, on several other factors among which are:

- The irrigation regime, especially frequency.
- Amount of applied water.
- Method of irrigation.
- Depth of water table.
- Hydraulic conductivity.

- Drainage efficiency.
- Climatic conditions.
- Soil texture.
- Impermeable layers in the soil profile and their depth.
- Kind of vegetative cover.
- Land topography.

To quantitatively predict the soil salinity as a result of irrigation with a particular water, the salt balance may be calculated.

Ayers and Westcot[6] calculated the average soil-water salinity EC_{sw} of the root zone by taking the average EC_{sw} of the soil surface, bottom of the upper quarter of the root zone, EC_{sw1}; bottom of the second quarter depth, EC_{sw2}; bottom of the third quarter depth, EC_{sw3}; and bottom of the fourth quarter depth which is the soil water draining from the root zone, EC_{sw4} or EC_{dw}. They suggested that:

$$EC_{dw} = EC_{sw} = EC_w/LF \qquad (17)$$

can be used to predict average soil-water salinity EC_{sw} in the rooting depth, knowing that EC_e (the salinity of the soil paste extract) is about half of the salinity of the soil water. Ayers and Westcot[6] came up with Table 5, which gives the concentration factor (x) for predicting soil salinity EC_e from irrigation water salinity (EC_w) and the leaching fraction (LF). Thus,

$$EC_e \ (dS \ m^{-1}) = EC_w \ (dS \ m^{-1}).x \qquad (18)$$

Evidently, EC_{sw} should decrease with the increase in the value of LF which is clear in Equation 17.

Assuming a steady state, conservation of mass, and that the plant absorbs 40, 30, 20, and 10% of its water respectively from the successively deeper quarter fractions of the root zone, Rhoades,[41] based on Rhoades and Merril[42] report in FAO[39] consultation on prognosis of salinity and alkalinity, evaluated the soil salinity after irrigation with water having a known salt concentration. He introduced an apparent leaching fraction La. It describes the fraction of infiltrated water passing a specified depth within the root zone (Rhoades and Merril[42]) as:

$$La = 1 - V_c/V_i \qquad (19)$$

where V_c is the volume of water consumed by evapotranspiration above a specified depth and equals (1–La)/(1–LF total) multiplied by the applied amount of water. LF total represents the LF obtained at the bottom of the root zone or profile, and V_i is the volume of water infiltrated. He concluded that at any depth, the concentration of the soil water will be $(C_i/La)C_i$. In terms of EC, $(C_i/La) \ Q_i$ where i refers to the irrigation water.

The concentration factor,

$$Fc = (1/La) \qquad (20)$$

can be used for the infiltrated water at a specified depth in the root zone, Table 5.

Ayers and Westcot[6] followed Rhoades'[42] approach to evaluate the soil salinity after being irrigated with water having a particular salinity expressed in EC units.

Table 5.   Relative solute concentration of soil water (field capacity basis) compared to that of irrigation water ($f_c = 1/L_a$) by depth in root zone and leaching fraction (L).[1]

Root Zone Depth in Quarters	V'_{cu}[2]	Leaching Fraction					
		0.05	0.10	0.20	0.30	0.40	0.50
0	0	1.00	1.00	1.00	1.00	1.00	1.00
1	40	1.61	1.56	1.47	1.39	1.32	1.25
2	70	3.03	2.70	2.27	1.96	1.72	1.54
3	90	7.14	5.26	3.57	2.70	2.17	1.82
4	100	2.00	1.00	5.00	3.33	2.50	2.00

[1]   Assuming 40:30:20:10 water uptake pattern in root zone.
[2]   Accumulative volume of consumptive use above this depth in root zone.
Source: Rhoades.[41]

The Salt-Affected Soils in Soil Classification Systems

The FAO/UNESCO[15] Soil Map of the World System

In this system of soil classification, the salt affected soils are grouped under:

- Solonchak: the main soil units are gleyic, takyric, mollic, and orthic. They are characterized by their high salinity within 125 cm of the surface. Salinity is high if it is more than 4 dS m^{-1} within the top 25 cm, or if it is more than 15 dS m^{-1} at some time of the year within 125 cm of the surface in coarse-textured soil surface, 90 cm in medium textures, or within 75 cm in fine textures.
- Solonetz: units are gleyic and orthic solonetz. They are characterized by the presence of a natric B horizon in which ESP is above 15 within the upper 40 cm of the horizon, or more exchangeable Mg+Na than Ca (at pH 8.2) within the upper 40 cm of the horizon if the saturation with exchangeable Na is more than 15% in some subhorizons within 2 m of the surface.

For soils without a natric horizon but having exchangeable Na more than 6% in a horizon within 100 cm of the surface, an alkaline phase is used.

The Russian Classification System

Solonchak soils contain a large quantity of easily soluble toxic salts in the upper horizon (30 cm) and have a total quantity of salts usually more than 2%. The pH of the soil solution is slightly alkaline (7.5).

The solonetz soils are characterized with high ESP in B horizon. The Russian system classifies these soils as presented in Table 6.

The Salt-Affected Soils in the U.S. System (U.S. Soil Survey Staff[43])

Though the salt-affected soils may be found in variable soil orders, soils of the arid regions are usually subject to salt accumulations. Accordingly, the salt-affected soils in the U.S.-system are mainly under Aridisols. It is the order of soils of arid regions charac-

Table 6.   Classification of solonetzic soils.

Degree of Alkalinity	ESP	
	Chernozem	Chestnut Brown
Solonetz	30	16
Strongly solonetzic	15–30	10–16
Medium solonetzic	10–15	5–10
Weakly solonetzic	10	5

Source: Kovda.[1]

terized with a light-colored surface, one or more horizons from clay (Argilic), gypsum (Gypsic), or calcium carbonate (Calcic). This order contains two suborders:

Suborder Orthids

Suborder orthids does not contain a clay horizon. It may contain a layer or horizon of gypsum, or salt as the group salorthids.

Suborder Argids

The surface layer has a light color followed by a clay layer rich in sodium thus called Natrargids, or a hard pan thus called Duraargids, or both thus called Nadurargids.

EFFECT OF SALINITY ON PLANT GROWTH AND COMPOSITION

The deleterious effect of salinity on plants has been demonstrated by several investigators. However, several factors may enhance or slow the salt effects on plants.

Two major groups of plants are known: nonhalophytic and halophytic.

The nonhalophytes are plants that can grow and flourish only in nonsaline media. Their growth in saline media is greatly affected.

The halophytes may be divided into:

- Salt-tolerant plants which can grow in nonsaline media and can tolerate concentrations of salts in their growth media.
- Plants that grow and flourish in the presence of salt concentrations. Their growth in nonsaline media is generally unsatisfactory. These plants have the ability to absorb many salts. Their cell sap usually contains chlorides more than other plants, though they absorb limited amounts of sulfates. The absorbed salts raise the osmotic pressure of the plants' cell sap to be higher than the osmotic pressure of the growth media. Transpiration—over time—may increase the cell sap salt concentration to a harmful level, though the halophytes usually avoid this increase by several mechanisms (Walter[44]):

Excretion

Few plants among which are grasses, bushes, and trees can excrete salts.

Regulation

Halophytes which cannot excrete salts can control the increase in salt concentration in their cell sap by absorbing and storing more water. Thus, they become succulent over time. A mechanism to store water is to decrease transpiration by closing the stomata during water stress (Radin[45]). In this regard, Fiscus[46] concluded that in most plant systems the stomata control the loss of water from the plant. However, it is the rate of water transport to the leaves that controls, to a large extent, the stomata.

Accumulation

Salts accumulate in the plant until they reach a certain concentration at which the plant dies.

Salts have the ability to enter plant tissues through the root with the other absorbed ions. The plants utilize the essential ions in their various processes. The other ions which are not needed, such as Nas, remain free in the plant sap and interfere in the biological processes inside the cells. Their concentration is increased with transpiration; unless the plant is capable of getting rid of these excess salts by excretion, regulation, or accumulation, it dies (Repp[47]). Repp adds that salt tolerance of plants' plasma results in:

- More salts that can be stored in the cells.
- Higher cell sap osmotic pressure and active suction pressure.
- Less harmful effects of salts when the cell loses water.

Aside from the halophytic plants, the factors which may enhance or slow the salinity effect on plants are related to the climatic conditions, the growth medium and the growing plant.

Climatic Conditions

The effect of salinity on growing plants under variable climatic conditions has long been demonstrated. Magistad et al.[48] grew three species of plants (beans, beets, and onions) in large outdoor sand cultures supplied with complete nutrient solutions having varying salinities. The salinity level at which a yield reduction of 25% of the control had taken place was determined graphically. The experiment was carried out in a cool and a hot location. The results are presented in Tables 7 and 8. The plant species studied showed greater tolerance to salinity in the cool than in the hot environment. Also, their order of tolerance to salinity had changed from onions > beets > beans at the cool location to beets > onions > beans at the hot location.

The same trend took place when plants were grown in hot dry seasons as compared with the same plants grown in cool humid seasons. In the hot dry environment, the plants suffer very high ET demand. The water absorption by the plant roots may be too rapid to compensate the soil water depletion and increased salt concentration around the roots. In the cool environments, water availability is not substantially affected. According to Ayers and Westcot,[6] climate appears to affect salt-sensitive crops to a much greater extent than salt-tolerant crops.

Table 7.   Response of three crops to salinity in sand cultures at two locations.

Crop	Salinity at Which 25% Yield Reduction was Observed (dS m⁻¹)	
	Cool Location	Hot Location
Bean pods	4.0	3.0
Garden beets	11.1	6.6
Onion bulbs	12.5	3.3

Source: Magistad et al.[48]

Kind of Crops

According to Ayers and Westcot,[6] there is an approximate 10-fold range in salt tolerance of agricultural crops. For example, production of faba beans is seriously decreased with a salinity level of 2 dS m⁻¹, while production of cotton is not affected. Tolerance of variable crops to salinity levels has been under investigation in numerous research centers. Crop tolerance tables for different field forage-, vegetable-, and tree-crops were assembled by the U.S. Sal. Lab.,[26] renewed by Bernstein[50] and by the University of California (USA) Commission of Consultants[51] in 1974, and updated by Maas and Hoffman.[52]

It should be pointed out that actual plant response to salinity varies with growing conditions as stated above. Thus, salt tolerance data cannot indicate accurate quantitative crop yield losses from salinity. Such data, however, indicate what might be expected when a selected crop might fare relative to another under similar conditions of salinity.

Crop Varieties

Varieties of a crop may differ in their salt tolerance. Varieties of rice had received considerable attention as rice is a main crop in salt-affected soils. Pearson et al.[53] compared the salt tolerance of 14 varieties of rice using percent germination and dry weight of seedling shoots. They concluded that growth of Bluebonnet, Caloro, Zenith, Agami Montakhab I, Asahi No. 1, and Kala Rata varieties was reduced at least 75% at weighted mean electrical conductivity of 5.1 dS m⁻¹. Kala Rata was the least tolerant of the 14 varieties tested during the germination stage, and the most tolerant of the six varieties tested during the young seedling stage. Agami Montakhab I was relatively less tolerant than some other varieties during the germination stage, but relatively more tolerant than most of the six varieties tested during seedling stage.

Researchers have been trying to make use of varieties of crops known for their tolerance to high salt concentrations in their growth media, even those considered wild or uneconomical, through hybridization with economical varieties to obtain new high-yielding, salt-tolerant crop varieties.

Stage of Growth

According to the U.S. Sal. Lab.,[26] crop species which are salt tolerant during later stages of growth, may be quite sensitive to salinity during germination. Sugar beets and safflower are sensitive during germination. Rice can tolerate a high concentration of salts

Table 8.   Effect of season on the relative rice yields.

Salinity of Root Zone (dS m⁻¹)	Relative Yield	
Approximate Range	Wet Season	Dry Season
Control (nonsaline)	100	100
2–4	93	81
4–8	63	53
10–12	39	11

Relative yields are comparable only within the same season. *Source:* Murthy and Jangrdhan,[49] cited in Abrol et al.[31]

at germination (up to 30 dS m⁻¹) but it is sensitive to salinity in the early stages of growth. Its tolerance increases with age during the tilling stage (Pearson[54]; Kaddah and Fakhry[55]). Barley, wheat, and corn follow a tolerance pattern similar to that of rice, while soybean tolerance to salinity depends on its variety (Maas and Hoffman[52]).

The Canada Department of Agriculture as stated by Abrol et al.[31] obtained the results presented in Table 10 which differ from those stated by Maas and Hoffman.[52]

Walter[44] relates this sensitivity to which the germ of the sensitive plants does not contain high concentrations of salts in spite of the fact that it is formed in the late stages of plant growth when the salt content of plants is at its maximum. Also, the osmotic pressure of the germ cell sap is usually lower than that of the plant cell sap and of the soil solution.

In the opinion of Ayers et al., the greater sensitivity of crops to salinity during germination is not certain. They showed that the degree of salinity that decreased the germination to 50% was the same degree that decreased the crop yield 50%, which indicated that the germination stage of growth is not more sensitive than other stages. The higher salt concentration caused by soil moisture evaporation in the upper soil surface layer, where seeds are planted, may be a cause of seed germination failure.

Salt Tolerance of Rootstocks

Fruit tree crops of one variety may differ in their salt tolerance according to rootstocks. Cerda et al.[56] showed that mandarin as a stock irrigated with saline water was characterized by its ability to exclude chloride ions, while the sour orange and Troyer citrange varieties in general accumulate high amounts of chloride ions in their leaves. In Bernstein's[57] opinion for many fruit crops, damage to plants could be related to the concentration of specific ions (Na^+ or Cl^-) rather than to the total soil salinity. He classified fruit crops with respect to specific salinity according to rootstocks (Table 9).

Effect of Salinity on Plant Absorption of Nutrients

Competition Between Ions

Ions in a solution affect their absorption of one another by a plant. In saline soils the situation becomes complex due to the pressure of various cation and anion species having different charges and hydrated radii. Also, inherent preferences exhibited by the absorption system are of importance. In this regard, potassium is usually absorbed more rapidly than other alkali cations which have smaller hydrated ions.

Table 9.   Tolerance of fruit varieties and rootstocks to chloride levels.

Crop	Rootstock/Variety	Limit of Tolerance to Chloride in Soil Saturation Extract mmol/l
Rootstocks		
Citrus (Citrus spp.)	Rangpur lime, Cleopatro	25
	mandarin, rough lemon	15
	tangelo, sour orange, sweet	10
	orange, citrange	
Stone fruit (Prunus	Marieanna lovell,	25
spp.)	Shalil Yunnan	10
		7
Avocado (Persea	West Indian,	8
americana)	Mexican	5
Varieties		
Grape (Vitis spp.)	Thompson seedless, perlette	20
	Cardinal, Blackrose	10
Berries (Rubus spp.)	Boysenberry,	10
	Olallie blackberry,	10
	Indian summer raspberry	5
Strawberry (Fragaria	lassen	8
spp.)	Shasta	5

Source: Bernstein.[50]

Table 10.   Tolerance of crops to salts at two stages of growth (Canada Department of Agriculture, 1977).

Crop	Germination Stage	Established Stage
Barley	Very good	Good
Maize	Good	Poor
Wheat	Fairly good	Fair
Alfalfa	Poor	Good
Sugarbeet	Very poor	Good
Beans	Very poor	Very poor

Source: Abrol et al.[31]

Competition between the alkali cations for absorption is a well known phenomenon. The uptake of one is generally reduced when the concentration of another is increased. The effect of Na^+ on the absorption of K^+ by plants has been reported by several investigators (Gauch and Wadleigh[58]; Shukla and Mukhi[59]; Devitt et al.[60]). On the other hand, application of K to saline soils decreases the deleterious effect of Na on plants (El-Fakharany[61]). Also, uptake of N as NH_4 was decreased with the increase of K application (Balba and Shabanah[62]).

Competition among anions also takes place. Zhukovskaya[63] reported that chlorides retarded the phosphate absorption and the sulphate inhibited uptake and accumulation of P by barley and sunflower. Torres and Bingham[64] showed that excessive NaCl in the substrate restricted NO_3^- uptake to the extent that N deficiency had taken place.

Effect of Salts on Plant Processes

Effect on Osmotic Pressure

Plants exert energy to extract water from the soil (the soil water potential). This energy should be greater than that with which the soil retains water. Salts in the soil solution increase the force the plant requires to be able to extract water (the osmotic potential). Thus in saline soils, the energy required for water absorption by plants should exceed the soil water potential plus the osmotic potential. The plant might suffer from reduction in absorbable water.

Magistad et al.[48] and Gauch and Wadleigh[58] studied the growth response of numerous crops in sand cultures in which relatively large quantities of chloride and sulfate salts were added to a control nutrient solution. Growth inhibition accompanying increasing concentration of added salts was virtually linear with increase in osmotic pressure, and was largely independent of whether the added salts were chlorides or sulfates. The slope of the negative regressions of yield on osmotic pressure of the substrate varied with salt tolerance of a given crop. Bernstein[65] stated that the presence of excessive concentrations of soluble salts in the root of plant may affect plant growth in a number of ways:

- The increased osmotic pressure of saline soil solution tends to restrict the uptake of water by plant roots.
- On the other hand, when the plants absorb the constituent ions of a saline solution, this may result, in some cases, in toxic accumulation of a particular ion or decreased absorption of some essential elements.

Later it was shown that plants can raise the osmotic pressure of their cell sap, meaning that decreases in water uptake and subsequent transpiration is not due to the higher osmotic pressure of the soil solution. Janes[66] showed that when bean and pepper plants grown in nutrient solution and subjected to water stress by gradual addition of NaCl or polythylene glycol (PEG) to increase the osmotic pressure of the nutrient solution, with each increase in the osmotic pressure of the solution, there was a proportional but smaller increase in the osmotic pressure of the expressed juice of the plants. Bernstein[67] stated that increased osmotic pressure of the cell sap adversely affects plant growth due to the following:

- The plants have to expend more energy to grow and maintain the higher osmotic pressure of sap.
- Cells such as plastids and mitochondria cannot change the osmotic pressure of their sap.
- The highest osmotic pressure might be accompanied with harmful concentrations of elements which might affect enzymes.

Effect of the Complementary Anion or Cation

Absorption of potassium by barley plants grown in sand culture and a nutrient solution varied according to the complementary anion of the applied K with $KNO_3 > KCl > KH_2PO_4 > K_2SO_4$ (Balba and Balba[68]). They related differences in K absorption as due to one or more of the following:

- Differences in the diffusion coefficients of the anions. Differences in the active concentrations of the salts due to variations in mean activity coefficients and the formation of ion pairs especially with SO_4^-.
- Lowering the pH of the medium in the case of KH_2PO_4 which may result in competition between H^+ and K^+.

Balba and Shabana[69] found more K absorption by barley plants when the nitrogen was applied as NO_3 than as NH_4^+. Balba and Soliman[70] showed that Na as NaCl decreased absorbed K by sudangrass more than Na_2SO_4. Also, the effect of anion depends on the complementary cation. The effect of NaCl on the potassium percentage in the whole bean plants was 4 and 6.5 times that with $CaCl_2$ or $MgCl_2$, respectively (Balba and El-Etriby[71]).

Effect of Salinity on Transpiration

Transpiration "ET" is a complex process in which several biological, physical, and climatological factors are involved. It is affected by the characteristics of both the growth media and the growing plants. The water flow from the plant to the atmosphere is controlled by the atmospheric evaporative demand which depends on several climatological factors such as temperature, relative humidity, wind velocity, sunshine, solar radiation, and others as stated above. Several investigators have shown that transpiration decreases with the increase in the salinity level of the growth media. This decrease was explained as due to the inability of plants to absorb water from a salt solution of higher osmotic pressure than that of the cell sap.

Myagkova[72] related the decrease of transpiration to the increase of the suction force of millet leaves grown on saline soils and the increase in bound water content. Lunin et al.[73] concluded that growth reduction with salinity resulted primarily from the reduction in water availability to plants. Strogonov[74] showed the following:

- Transpiration rate of plants from SO_4^--salinized soils was markedly increased compared to the control.
- The corresponding transpiration rate of plants from Cl^--salinized soils was markedly decreased.
- The absorbing area of the roots in control and SO_4^- plants was essentially the same, while that of the Cl^- plants was much smaller.

Shalhavet and Bernstein[75] pointed out that by increasing soil salinity, the transpiration rate per unit leaf area was constant regardless of the salinity level.

Table 11 shows that the plants grown under saline conditions suffered from a decrease in their root system through which water is absorbed. Also, the surface area of their leaves was decreased and the leaves retained more water. Thus, transpiration of plants grown under saline conditions decreased. Thus, transpiration of plants growing in saline media decreases as a result of: (1) lower absorption of water due to adverse effects on roots and/or increase in the soil solution osmotic pressure, thus the rate of water transport to the leaves which control the stomata is slowed (Radin[45]), (2) decrease in the surface area of leaves, and (3) increase in water retained by the plants. It might be concluded that the salinity of the growth media hinders water absorption by plants due to adverse effects on roots and/or increase in the soil solution osmotic pressure. The net result is lowering the plant transpiration accompanied with decreased plant growth and yields.

Table 11.   Effect of different salts on sundangrass plants.

	Tap Water	NaCl	Na_2SO_4	$CaCl_2$
Root, dry weight, g/pot	62.2	48.4	41.5	30.5
Above growth, g/pot	18.9	13.7	11.6	9.0
Total leaf area, cm²/pot	2696	2018	1860	1615
Total ET, ml/pot	4651	2886	3772	3495
ET in last 10 days,				
ml/pot	2628	2147	2062	1850
mm/day	8.37	7.93	7.06	5.89
ET/l g dry weight	57.3	62.60	70.90	88.40
ET/cm² of leaf area	1.72	1.92	2.00	2.10

Source: Balba and Soliman.[70]

Table 12.   Effect of Na_2CO_3 on forms of soil Fe, Mn, Zn, P, and Ca.

		Untreated Soil	Na_2CO_3
Na_2EDTA Ext.	Fe, ppm	11.0	6.7
NH_4OAC	Fe, ppm	10.0	5.2
Easily reducible	Mn, ppm	20.0	14.0
NH_4OAC	Mn, ppm	1.5	1.0
NH_4OAC	Zn, ppm	0.5	0.38
NaHCO Ext.	Ca P, ppm	12.4	2.4
Soluble	Ca, meq/100 g	1.32	0.42
Exchangeable	Ca, meq/100 g	17.40	12.10

Source: Balba and El Khatib[76] and Thabet.[77]

Effect of Salts on Forms of Soil Nutrients

Table 12 shows results obtained by Balba and El-Khatib[76] and Thabet[77] by incubating 100 g soil for 15 days with water or Na_2CO_3 solution (2 meq/l). Extractable Fe, Mn, Zn, and P decreased due to their transformation to other forms by the Na_2CO_3 treatment (pH effect). Balba[78] showed that increments of NaCl or $CaCl_2$ enhanced the immobilization of the superphosphate applied to Egyptian alluvial soil. He explained that $CaCO_3$ solubility was increased in the presence of NaCl, bringing more Ca^{2+} into the solution. Thus, more phosphate precipitation may take place. The Ca^{++} of $CaCl_2$ has the same effect on soluble phosphate.

Chu and Chang[79] in Taiwan and Sinha and Bhattarcharya[80] in Bihar-India found low available P in the saline soils. Rankov[81] found low P and K in the solonetz soils of Bulgaria. Bains and Fireman[82] considered the soil fertility as a limiting factor under saline conditions.

Effect of Salts on Nutrient Loss from the Soil

El-Shakweer and Abdel Ghaffar[83] showed from the N balance sheet for cotton grown on 5 different soils that the gaseous loss of N was the highest from the nonsaline sodic and the lowest from the saline soils. Balba et al.[84] showed that amounts of K in the leachates of sandy soil pots planted to tomatoes and irrigated with tap water (EC=0.4 dS m⁻¹) and waters having EC = 2.8 and 5.4 dS m⁻¹ were 57, 110, and 264 mg/pot, respectively. Chhabra et al.[85] showed that leaching columns of a sodic soil caused substantial losses of P.

Effects of Salts on Soil Microorganisms and Their Activities

Several investigators have shown that the number of microorganisms in general in salt-affected soils is lower than in the nonsaline soils. Also, activities of the microorganisms under saline and/or sodic conditions are slower than under normal conditions (Paliwal and Maliwal[86]) (Figure 4). Effect of salinity (NaCl) on nodulation of legume plants depended on the kind of plant. Bernstein and Ogata[87] showed that nodulation of soybean plants variety Lee was strongly reduced with increasing salinity. On the other hand, nodulation of alfalfa was only slightly affected.

Constraints Related to the Soil Physical Conditions

Nonsaline sodic soils usually suffer from unfavorable physical properties. Low salinity and high ESP values result in dispersion of soil particles which cause a permeability problem. The poor soil permeability makes it more difficult to supply the crop with water and causes nutrition problems. Tilling practices of these soils are difficult.

Shallow water tables may reduce growth due to decrease of rooting volume and insufficient oxygen. Woodruff et al.[88] found that soils with water table 70 cm deep, required more N fertilization to reach maximum corn yields than soils with deeper water tables (Figures 5 and 6).

Toxicity and Specific Ion Effects

Certain elements (e.g., B, Li, Se, and other micronutrients) are, under specified conditions, toxic to plants in minute amounts. Other cations and anions, though not actually toxic, may have deleterious effects.

The Specific Effect of Salts on Plants

Uhvits[89] studied the effect of isosmotic solutions of NaCl and mannitol on alfalfa seeds. The results were clear that at equal osmotic pressures, NaCl depressed germination much more than mannitol. This was especially marked at high osmotic pressures, 9 atm or more. For instance, in a solution of NaCl of osmotic pressure of 15 atm 7, germination after 10 days was 2%, while in an isosmotic solution of mannitol it was 57%.

Several explanations were offered for the specific effect. (1) The ions penetrating the cell have a specific effect. Cations having coagulation properties decrease the permeability of the protoplasm to salts, while ions having a peptidizing effect on plasmocolloids, increase the permeability to salts. Thus, the ability of neutral salts to penetrate the protoplasm is the result of the opposing colloidochemical properties of the ions. The toxicity of salts is directly correlated to their permeability. The more rapidly salts penetrate and

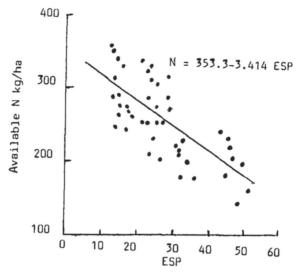

Figure 4.   Relation between ESP and available N of sodic soils (Paliwal and Maliwal,[86] 1972).

accumulate, the more toxic they are to the plants (Strogonov[74]). (2) Salts and/or exchangeable Na, which raise the pH of the medium, precipitate Ca, inducing Ca deficiency. Thus excessive Na^+ or CO_3^- uptake adversely affect the plant (Balba and El-Etriby[71]).

Toxicity of Na

In the nonsaline sodic soils (pH > 8.5, EC < 4 dS m^{-1}, and exchangeable Na > 15%) plants suffer from the excessive exchangeable Na in two manners. The first is due to its competition with other nutrient cations as stated above. The second is due to its specific deleterious effect. Plants differ in their sensitivity to exchangeable Na. The very sensitive plants do not tolerate exchangeable Na as low as 2–10%. On the other side, many field crops tolerate exchangeable Na as high as 40% of the exchange capacity. Pearson[54] and El-Gabaly[90] had shown that plants grew normally on soils containing ESP as high as 30–50.

Effect of CO₃

Faba beans irrigated with Na_2CO_3 solution ceased to grow after 30 days (Figure 7) (Balba and El-Etriby[71]). The bicarbonates are usually considered similar to the soluble carbonates as they are transformed to the latter. When the $CO_3 + HCO_3^-$ concentration in the irrigation water exceeds the Ca + Mg concentration, the quality of the water is lowered (Eaton[8]).

It was stated above that HCO_3^- and/or CO_3^- might induce transformation in the forms of soil nutrients as Fe and Mn. Several investigators had shown that bicarbonates induce iron chlorosis in plants due to disturbance in their nutrition. Hale and Wallace[91] have demonstrated that increasing both HCO_3^- and CO_3^- decreased accumulation of iron. They reported that under iron chlorosis, leaves were high in P and in P / Fe ratio and lower in silicon but higher in aluminum than green plants.

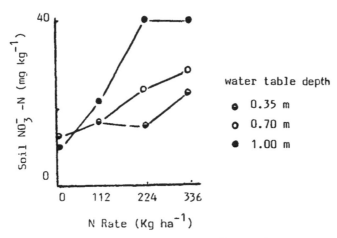

Figure 5.   Effect of water table depth and added N on soil.

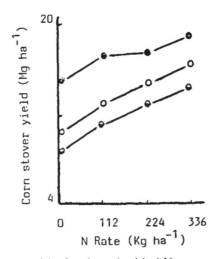

Figure 6.   Effect of water table depth and added N on corn yield.

Increase in HCO_3^- and/or CO_3^- in the soil solution and/ or irrigation water causes the precipitation of the soil Ca and Mg. Thus, deficiency of both cations is expected as in the case of nonsaline sodic soils. At the same time toxicity of Na is accentuated (Eaton[8]).

Chloride Toxicity

Chloride is not adsorbed by soils, but moves readily with the soil water. Leaf burn due to Cl toxicity usually starts at the leaf tip of the older leaves and progresses back along the edges as toxicity increases. Abnormal early leaf drop and defoliation may also take place (Ayers and Westcot[6]).

According to Rhoades and Loveday,[92] attention must be paid to both Na and Cl concentrations of the irrigation water if sprinkler irrigation is used. Tolerance to Cl and Na+ concentrations in the irrigation water is markedly reduced as the plant absorbs salts through its leaves.

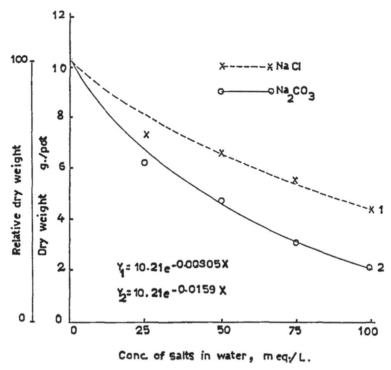

Figure 7.   Effect of NaCl and Na_2CO_3 waters on the dry weight of faba beans 30 days old (Balba and El-Etriby[71]).

Boron Toxicity

Though B is an essential nutrient needed in small amounts, excessive B in the growth medium is toxic. Toxicity symptoms show first on leaf tips and edges as yellowing, spotting, or drying of leaf tissues. On trees of almond seriously affected, gummosis or exudate on limbs or trunk is noticeable (Ayers and Westcot[6]).

Boron is present in the soil in the adsorbed and soluble forms. Thus, it requires an excessive amount of water to reach the soil (Figure 8). Plants respond primarily to the boron concentration of the soil water rather than to the amount of adsorbed B. The potential of creating a boron problem upon irrigation is assessed by comparing average boron concentration in the soil water with levels tolerable without yield reduction for the crop(s) in question.

The Content of Soil Water

The available soil water ranges from the field capacity to the permanent wilting point (PWP). The salt concentration in the soil water increases with the decrease of this water by evaporation and transpiration. At PWP, the salt concentration is about double its concentration at the field capacity. Accordingly, the adverse effect of salts is expected to increase. Also, the soil water potential increases, making it more difficult for the plant to absorb its need of water. Thus, as the available soil water decreases, between irrigations or when water is short, the water deficit and osmotic effects become greater. Crops vary

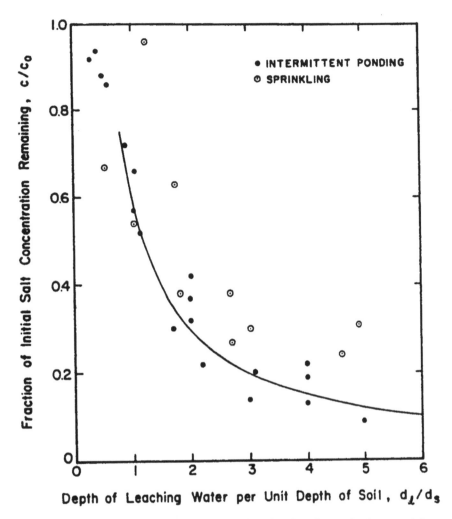

Figure 8.   Depth of leaching water per unit depth of soil required to reclaim a soil inherently high in boron (Rhoades, 1980).

in their water requirements, optimum growth season, rooting depth, and cultural requirements. Soil salinity is affected by soil and water practices suitable for various crop rotations. Crops which require flooding or more frequent irrigations reduce soil salinity, while others which require a long period without irrigation in hot dry seasons before harvesting will increase soil salinity.

The effect of soil and water cultural practices on soil will be discussed later.

Quantifying Plant-Salt Relation

Under saline conditions, the plant growth is a function of soil and/or water salinity. The quantitative expression of the effect of salts on plant growth would allow the calculation of the expected yield from knowing the salinity level. The U.S. Salinity Laboratory[26] expressed the effect of salts on plant production by evaluating the salt concentration which decreased the yield of a crop to 50% of its yield under nonsaline conditions.

Mass and Hoffman[52] expressed the relative yield (Y) at any given soil salinity (EC_e) by the equation:

$$Y = \frac{100 \ (EC_o) - EC_e)}{EC_o - EC_{100}} \qquad (21)$$

where EC_{100} is the salinity threshold value (EC_e where $Y = 100\%$) and EC_o is the salinity at zero yield (EC_e where $Y = 0$).

Robinson[93] presented straight line equations to predict yields of onion, wheat, beans, cabbage, carrots, and alfalfa when sprinkler-irrigated with the Colorado River water having $EC = 1.35$ dS m^{-1} or 2.0 dS m^{-1}.

de Wit[94] tried to relate dry matter yield "Y" to transpiration on the basis that the influence of soil water on transpiration is the same as on yield. Therefore, if transpiration can be measured, the yield can be estimated. The equation suggested by de Wit[94] appears as follows:

$$Y = mT \ / \ E_o \qquad (22)$$

where T is transpiration in cm, E_o is average free water evaporation rate in cm day^{-1}, and "m" is a crop factor having dimensions of kg ha^{-1} day^{-1} if yield is in kg ha^{-1}. Hanks[95] found out that de Wit's equation holds for conditions of differential stata. With "m" and "E_o" being constant, i.e., for one crop and year, the relation of relative transpiration to relative yield can be expressed as follows:

$$Y \ / \ Y_p = T \ / \ T_p \qquad (23)$$

where "Y_p" is potential yield when transpiration is equal to potential transpiration. T_p is defined as the transpiration which occurs when soil water does not limit transpiration.

Childs and Hanks[96] tried to develop a model relating the factors influencing crop growth and salinity for predicting the long-term effects of different irrigation management schemes. They used a model previously introduced by Nimah and Hanks[97] which included a "root extraction" term related to the water potential gradient. The plant root extraction term A (z,t) is defined as follows:

$$A \ (z, t) = \frac{[H_{root} + (RRES.z) - h(z,t)] - S(z,t) \ . \ RDE(z) \ . \ K(Q)}{X \ . \ Z} \qquad (24)$$

where H_{root} is an effective water potential in the root at the soil surface where z is considered zero and RRES is a root resistance term equal to $(1+RC)$. RC is a flow coefficient in the plant root system assumed to be 0.05. When RRES is multiplied by z, the product will account for the gravity term and friction loss in the root water potential. Thus, the root water potential at depth z is higher than the root water potential at the surface (H_{root}) by a gravity term and friction loss term (assuming that the friction loss in the root is independent of flow). h(z,t) is the soil pressure head, S(z,t) is the salt (osmotic) potential (in equivalent head units). RDF(z) is the proportion of total active roots in depth increment z, K(Q) is the hydraulic conductivity at depth z, and X is the distance between the plant roots at the point in the soil where h(z,t) and S(z,t) are measured. X is arbitrarily assumed to be one. Dividing by Z converts the transpiration flux into change of water content per unit time.

Nimah and Hanks[97] stated that their model predicted significant changes in root extraction, evapotranspiration, and drainage due to the variations in pressure head-water content relations and root depth. Variations in the limiting root water potential had a small influence on estimated evapotranspiration, drainage, and root extraction.

Plants respond differently to different salts, but the quantitative approach to salinity of soil and water was mostly dealt with without considering the salt composition and the specification effect. Szabolss[98] reported that the relative effect of Na_2SO_4 : $NaHCO_3$: $NaCl$: Na_2CO_3 was in the order of 1 : 3 : 3 : 10.

Using straight line equation and the slope ratio method, Balba and El-Etriby[71] showed that at the age of 30 days, the harmful effect of Na_2CO_3 to faba bean plants grown in pots was twice as much as that of $NaCl$. At the age of 60 days, the relative effects of $NaCl$: $NaSO_4$: $CaCl_2$: $MgCl_2$: $(NaCl+CaCl_2+MgCl_2)$ were equal to 1 : 0.84 : 0.9 : 0.92 : 0.75. The salt concentration which decreased the plants dry weight to 50% at the age of 30 days was 85.5 meq/l and at the age of 60 days was 51.7 meq/l in the case of single salts and 89.7 meq/l in the case of the triple salt mixture.

Management of Salt-Affected Soil

Reclamation

General

In arid ecosystems, rain is scanty. Utilization of soils in this ecosystem depends on availability of a dependable satisfactory source of water, otherwise the lands would remain barren. Accordingly, management practices of these areas depend mainly on water management. Water is essential for the reclamation processes of the land as well as for supplying the plants and animals with their need of water. Also, draining the excess water during the reclamation or the utilization stages is a decisive process. Lack of an efficient drainage system would hinder the reclamation process and cause waterlogging.

Selection of suitable crops, tillage and its suitable machinery, and other practices come after guaranteeing a dependable source of irrigation water. A main characteristic among all activities in these lands is to counteract the effect of aridity and lighten the adverse effect of salinity which is—in turn—a result of aridity.

Reclamation of the salt-affected soils requires (Rhoades and Loveday[92]):

a. Removal of excess salts from the root-system zone or deeper in the soil, to a level tolerable by the selected crops.
b. Lowering the groundwater table beyond the root zone.
c. Neutralization of Na_2CO_3 and replacement of the exchangeable sodium by calcium.
d. Protecting the soil from resalinization.

Preparatory Steps

Before starting the reclamation processes several preparatory steps should be taken:

1. Soil survey and mapping.
2. Examination of available water resources.

3. The land of the project is divided on the soil map into parts by canals, drains, and roads.
4. The layout on the map is carried out on the ground starting with the borders of the large divisions (main canals and drains).
5. Locations of constructions are selected.
6. From the contour map, amounts of earth moving required for leveling and grading in each part is estimated.
7. The needed suitable machines and their requirements are estimated.
8. Knowing the levels of the water source and the land after leveling, the need for lifting water will be known. The method of irrigation is selected to match the type of the soil and the kinds of crops to be grown as well as the method of leaching the soil.
9. Knowing the depth of the water table and the highest level of the outlet of the drainage system, and knowing the amount of water required for leaching the salts, the amount of the drainage water and its level and the required lifting can be discovered.
10. The needed personnel, their fields of specialization, and their levels of qualifications are pointed out.

Prereclamation Processes

The soil survey mentioned above should describe the following:

*   Permeability of the horizons of the profile.
*   Characteristics of the water table: its depth, fluctuations, and salt concentration and composition.
*   Characteristics of the irrigation water: quantity, quality, its surface level relative to the soil level, and time of availability.
*   Land topography (topographic map) relief and drainage outlets.
*   Climatic characteristics of the region: rainfall, temperature, and evaporation.

The above-stated information is used to calculate and carry out the following:

*   Amount of water required for leaching.
*   The necessary leveling.
*   Layout of canals, roads, and constructions.
*   Type of drainage and layout of drains.
*   Expected amounts of drainage water.

Reclamation Steps

1. Separation of the land to be reclaimed from neighboring lakes, swamps, or canals which have a higher water level than the land by a running drain.
2. Separation of the area from elevated neighboring areas by running drains.
3. An efficient drainage system is as important as an adequate source of suitable water.
4. Tillage enhances leaching the saline soil salts as it loosens the dense subsoil. It is advantageous to deep-plow the soil during the reclamation stage.
5. Land leveling is of special importance for efficient reclamation. The field should be sufficiently flat to achieve uniform water distribution. Salts usually accumulate

in high spots of the field, while in low-lying spots water accumulates and causes waterlogging. Failure of seed germination can occur in both cases: high-saline spots and low-lying waterlogged spots.

6. Grading is carried out after establishing the canals, drains, and roads according to the layout of the land to be reclaimed. Leveling is performed according to the contour and soil survey maps. Scraping the surface layer for leveling exposes the subsurface layer. Fertility problems may arise because of the low fertility of the subsurface soil layer. Also, scraping the surface layer of shallow soil profiles may cause the water table to be closer to the soil surface. Leveling may be carried out for the whole area or after dividing the area into parts. Each part is graded according to its own level and separated from the other parts differing in their levels by drains.

After leveling, the canals and drains are dug according to their layout.
Fine leveling follows.

The Leaching Process

The leaching process signifies infiltration of an amount of water through the soil profile for dissolving and carrying the salts with the applied water deep in the soil far from the rooting zone or to the drainage system. The process should achieve the following:

- Dissolution of soil soluble salts and removing them from the soil profile.
- Removing the sodium salts formed as a result of replacing exchangeable Na with Ca in reclaiming the soil sodicity.

The removal of salts is carried out by leaching. Scraping the salt crust and/or flushing the saline soil surface with water are practiced by some farmers, but they do not replace leaching. The leaching process is affected by the following:

a) The amount of water available for leaching.
b) Concentration and composition of salts in the water and in the soil.
c) The soil permeability.
d) Depth of groundwater and concentration and composition of its salt content.
e) The efficiency of the drainage system.
f) Method of water application.
g) Type of leaching.

The work carried out by Balba[99] indicated that the soil texture is of special importance in the leaching process. The salt concentration of the water is important if the soil salt content is low. In highly saline soils this factor is of less importance.

Mechanism of Salt Removal by Leaching Salt-Affected Soils

When water infiltrates through the soil profile from its surface downward, several transformations in the soil constituents take place (Table 13).

- All soluble salts are dissolved as long as amount of the applied water is enough. The dissolved amounts of the sparingly soluble salts such as calcium carbonate increase.

Table 13.   Variation of the soluble cations percentages after leaching the soil column with 300 ml of water.

	Column Layer			Total Column	Total Initial
	1st	2nd	3rd		
Ca % of total cations	51.6	25.6	17.4	26.8	7.8
Na % of total cations	19.4	45.1	52.2	41.6	67.1
Mg % of total cations	29.1	29.2	30.4	31.6	25.1

Source: Balba and El-Laithy.[130]

- The dissolved salts move with water toward the bottom of the soil profile and to the drains.
- The equilibrium condition among the soluble and exchangeable cations is changed. Also, the ratios between the exchangeable cations are changed.
- Precipitation also may take place as the precipitation of Ca in the presence of excess Na_2SO_4 and the formation of $CaCO_3$ in the presence of Na_2CO_3.

The chromatography theory is used to demonstrate the salt movement in the soil when water is applied. The water fills the soil pores, forming a thin layer on the soil particles surfaces. Increasing the water rather than the volume of pores, the water moves downward carrying the dissolved salts from the upper layer to a lower one. Again the pores of this layer are filled and the water moves to a lower layer, and so on, until the end of the soil column where it infiltrates with its load of salts (Gardner and Brooks[100]).

Using sandy soil columns salinized with Na labeled with ^{22}Na, Balba and Bassiuny[101] traced the NaCl movement with increments of water. They showed that upon application of an amount of water equal to the soil water-holding capacity, the salts were accumulated in the lower fourth of the soil column. After an application of a larger amount of water than the soil's water-holding capacity, the extra water filtrated out of the soil column, and the soil started to lose its salts.

Methods of Leaching

Flooding and Sprinkling

Two methods of water application may be used, namely flooding and sprinkling. Experimental results have shown that leaching by water when the soil moisture content is less than saturation is more effective than when the soil is saturated with water. An unsaturation condition is obtained by intermittent ponding or by sprinkling at rates less than the infiltration rate of the soil. Nielsen et al.[102] showed that 25 cm of sprinkled water reduced the salinity of the upper 60 cm of soil to the same degree as 75 cm of ponded water. Also, a smaller fraction of salts moves up during evaporation from the soil previously leached by sprinkling than from the same soil leached by ponding water.

Continuous and Intermittent Leaching (Figures 9 and 10)

Two methods are practiced in the leaching process by flooding: continuous and intermittent.

Continuous Leaching

In continuous leaching, the sufficient amount of water is ponded on the soil surface. The water is usually applied in portions provided the depth of water on the soil surface is of less than 10 cm. The soil is not allowed to dry until the required amount of water is applied. The soil is then plowed and sampled for EC determination. Continuous leaching gives good results under the following conditions:

- Soils having good permeability.
- Saline soil water table is close to the soil surface.
- Climatic conditions enhancing high evaporation rate.

Intermittent Leaching

In this method the required amount of water is divided into several applications. The soil is usually dried and plowed before each application. This method is practiced in soils having poor permeability. However, soils with a high water table usually are resalinized in the drying periods. In this regard, El-Gabaly[90] found that: a) the period between the successive water applications should not exceed 20 days in winter or 10 days in summer to prevent resalinization of the top soil, b) the final salinity after intermittent leaching was generally higher than after continuous leaching, and c) the time required to leach the top 60 cm was about 40% more in the intermittent than in the continuous leaching. These findings in general agree with Rhoades[103] who stated that continuous leaching by applying a depth of water equivalent to the depth of the soil to be reclaimed will remove about 70% of salts initially present in the soil. Through intermittent ponding of water or by sprinkling, one third of this amount of water was needed, though the time of reclamation was extended.

The Amount of Water Needed for Leaching Saline Soils

The amount of water available for leaching controls the area that can be leached. Evaluating the amount of water necessary for leaching per hectare is the first step in planning any reclamation project. This problem has been the concern of several soil scientists.

Farmers used to depend on their experience. They applied 7000–15000 m^3 per hectare to leach saline soils depending on the soil texture and permeability, the soil depth to be leached, and the final soil salinity level after leaching.

Among the attempts to evaluate the amount of water needed to leach the saline soils were those of Van der Molen,[104] Kovda,[105] Bryssine[106] depending on the salt balance, and Gardner and Brooks[100] depending on the chromatography theory in soil columns.

Using Gardner and Brooks'[100] mathematical model for the leaching process, Balba and Bassiuny[101] concluded that the ratio between the depth of added water and the depth of soil to which 80% of salts must be removed depended on the soil texture. Thus, this ratio in case of sand particles (0.2–0.4 mm) was 0.55, in sand particles (0.08–0.2 mm) the ratio was 0.65, while in case of the Nile deposits, the ratio was 0.9. Thus, if the depth of soil is 1 m; 55, 65, or 90 cm of water must be applied to remove 80% of the salts originally present from the surface to the depth of 1 m. Pearson and Ayers'[107] corresponding value is 1 m of water to leach 80% of salts from a depth of 1 m and to leach 90% of salts from the depth of 50 cm.

Several investigators have tried to develop models simulating salt movement and reactions in soils during leaching (Bresler et al.[108]; Oster and Halverson[109]; Tanji et al.[110]). In Rhoades'[103] opinion, simpler empirical relations can be recommended as a practical approach to estimate the amount of leaching water required for reclaiming saline soils. He stated that Jury et al.[111] in a large outdoor lysimeter column found that 1.5 pore volume (PV) of applied water, or 1 PV of drainage water, removed essentially all chloride salt from both sandy loam and clay loam soils by either ponded or unsaturated leaching. Dissolving soil gypsum and total salt removal required considerable additional leaching. The following curve described total salt removal:

$$(C/C_o)\ (d_l/d_s Q) = 0.8 \tag{25}$$

where C is salt concentration in effluent, CO is initial salt concentration of soil water, dl and d are depth of water applied and depth of soil, respectively, and Q is soil volumetric water content.

The term $(d_l/d_s Q)$ is equal to pore volumes of leaching water applied.

Rhoades[103] stated that about 60 and 80% total salt removal occurred with application of 2 and 4 pore volume equivalent of leaching, respectively. Saline soils containing excessive amounts of boron require special attention in leaching because boron is difficult to leach. The situation is more difficult as boron is toxic to plants when present in concentrations as low as 0.1 to 1.0 mol m^{-3}. Thus, leaching boron from soil requires more water than the amount required to leach other soluble salts. The amount of water required to remove a given fraction of boron is about twice that required to remove soluble salts by continuous ponding. Because minerals containing boron may release it by weathering, periodic leaching may be required (Bingham and Garber[112]; Rhoades[103]).

Leaching Saline Sodic Soils

In leaching saline sodic soil columns with water, three processes take place and are responsible for the displacement of the adsorbed sodium:

a) The hydrolysis of the Na-clay in water results in replacement of adsorbed sodium by another cation in the solution, especially divalent cations. If divalent cations were not present, mostly hydrogen ions would be retained in place of sodium (Wiklander[113]).

b) In soils containing Ca-salts of low solubility such as gypsum or calcium carbonate, increasing the amount of water increases the amount of Ca^{++} in solution. This effect is enhanced by the presence of neutral soluble non-calcic salts. Under these conditions calcium is usually the cation which replaces adsorbed sodium.

c) The hydrolysis of the soil CaCO$_3$ in water in the presence of Na-soil ends up with partial displacement of adsorbed Na by Ca^{++} from the soil CaCO$_3$ as follows (Bower and Goertzen[114]):

$$2Na\text{--ads} + CaCO_3 + 2H_2O \rightarrow Ca\text{--ads} + 2HCO_3 + 2NaOH$$

They stated that the equation shows that with the increase in water, the reaction proceeds to the right side owing to the removal of NaOH and HCO_3 until all the adsorbed sodium is replaced.

Under these conditions the Ca minerals solubility is enhanced five-fold or more (Oster et al.[109]).

The results of these reactions agree with Wiklander[113] who showed that dilution favors the adsorption of the divalent cations. Balba[115] also showed that when irrigating saline sodic soil with tap water in pots, its exchangeable Na percentage decreased, and the $CaCO_3$ percentage decreased from 2.2 to 0.5%. Dielman[116] obtained similar results using gypsiferous soil in Iraq and showed the decrease of its gypsum content after leaching.

Several investigators (Oster and Halverson[109]; Oster and Frankel[117]; Jury et al.,[111] as stated by Rhoades[103]) consider that excessive exchangeable sodium in saline sodic soils is sufficiently reduced during leaching and that no special reclamation procedures are required.

In practice, sodic soils without neutral soluble salts are usually dispersed. Their hydraulic conductivity is very low and the leaching is slow. In the opinion of this book's author, the contribution of the calcium of the soil $CaCO_3$ to the displacement reaction of the adsorbed Na is not effective unless neutral soluble salts are present. Under this latter condition the solubility of $CaCO_3$ increases and the pH of the system is slightly alkaline. In the work of Schoenover et al.,[117] application of water to fine-textured sodic soil having pH 9.6, and low in soluble salts, did not effect displacement of adsorbed sodium.

Reclamation of Nonsaline Sodic Soils

These soils suffer from several problems which adversely affect their utilization. These problems may be one or more of the following:

1. The poor physical properties described earlier in this chapter due to the dispersion of the soil particles.
2. The alkaline pH as a result of soil sodicity.
3. Many soil nutrient forms are unavailable to plants under such alkaline conditions.

In order to tackle these problems, the reclamation processes should aim at:

1. Replacing the excess exchangeable Na by Ca.
2. Efficient drainage system to prevent upward movement of groundwater rich in Na_2CO_3 or $NaHCO_3$ and to remove the sodium salts formed as a result of the reaction between the applied amendment and the soil.
3. Improvement of the soil physical properties.
4. Fertilization with the proper fertilizer and selection of crops tolerant to sodicity.

Reclamation of nonsaline sodic soils is carried out by replacing the excess exchangeable Na by Ca^{++} (U.S. Sal. Lab.[26] and Rhoades[103]). The source of Ca^{++} may be $CaCO_3$, $Ca(OH)_2$, $CaSO_4$, or $CaCl_2$. Acids may also be added to these soils on the basis that the acids dissolve soil-insoluble Ca-compounds either directly with H_2SO_4 or HCl or indirectly with the application of sulfur which is oxidized by the soil microorganisms to H_2SO_4. Application of organic matter also exerts organic acids and H_2CO_3. Iron sulfate or aluminum sulfate (and mixtures of lime and sulfur) have acidic reactions. The formed acids or the acidic reactions of salts dissolve the soil Ca-compounds. Selection among the materials used for reclamation depends on the sodic soils characteristics as follows:

1. Soils containing calcium or Mg carbonates.
2. Soils having pH of the surface layer below 7.5 and almost free of $CaCO_3$ such as the upper layer of the solonetz soils.
3. Soils having pH higher than 7.5 and not containing $CaCO_3$.

Generally, soils containing $CaCO_3$ can be reclaimed by any of the above stated materials which supply the soil directly or indirectly with Ca^{++}. However, because of the alkaline pH of the nonsaline sodic soil, application of $CaCO_3$ or $Ca(OH)_3$ is not recommended for the soils which contain originally $CaCO_3$.

Acids, acid-forming materials, or salts having acidic reaction are not recommended for reclaiming soils which do not contain $CaCO_3$. The reclamation in this case is carried out with the application of any calcium source. Selection among these sources depends on the Ca supplied and the cost of the material.

In the case of the solonetz soils characterized with their slightly acid $CaCO_3$-free surface layer, a B-horizon rich in exchangeable Na, and a C-horizon rich in $CaCO_3$, the sodicity is reclaimed by applying $CaCO_3$ or by deep plowing and turning over the $CaCO_3$-rich soil layer. The applied or soil $CaCO_3$ supplies Ca ions after being mixed with the acidic soil surface, exchangeable Na is replaced by Ca ions infiltrated from the above soil layer (Table 13).

Because of its physical properties, lower cost, and natural presence in arid regions, gypsum is the main Ca supplier for reclamation of sodic soils.

$$Na_2 - clay + CaSO_4 \rightarrow Ca - clay + Na_2SO_4 \qquad (a)$$

The Na_2SO_4 should be leached in order that the reaction proceeds to the right side.

In nonsaline sodic soils which usually contain considerable concentrations of Na_2CO_3 the following reaction takes place:

$$Na_2CO_3 + \quad \rightarrow CaCO_3 + Na_2SO_4 \qquad (b)$$

Both reactions "a" and "b" have extensively been studied by soil investigators in several regions, especially in the arid and semiarid regions. It has been demonstrated by laboratory—as well as field—studies that gypsum is the main source of Ca for reclamation of the sodic soils. Also it has been shown that gypsum has a beneficial effect on the soil structure, permeability and hydraulic conductivity (Abrol et al[31]). Its application supplies the infiltrating water with the electrolyte concentration required to maintain soil permeability. Rhoades[103] constructed a guideline figure (Figure 11) to show the relation between SAR in the topsoil and EC of infiltrating water and their effect on the soil permeability hazard. This beneficial effect is reflected on the efficiency of the reclamation process.

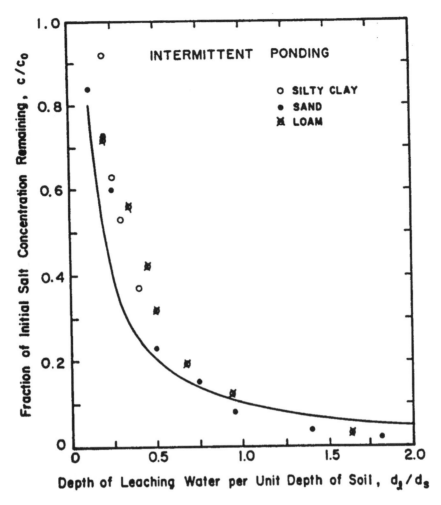

Figure 9.   Depth of leaching water per unit depth of soil required to reclaim a saline soil by ponding water intermittently (Rhoades, 1980).

Determination of the Amendment Requirement

Replacing the Soil Exchangeable Na by Ca

Amendments are applied to the soil to decrease the exchangeable sodium percentage (ESP) to less than 15. The process consists of the determination of:

1. The soil exchangeable sodium and ESP.
2. The amount of exchangeable Na that should be removed by the applied amendment:

$$\text{Initial exch. Na - Final exch. Na}$$

$$\text{or (ESP initial - ESP final). CEC, mole/100 g soil}$$

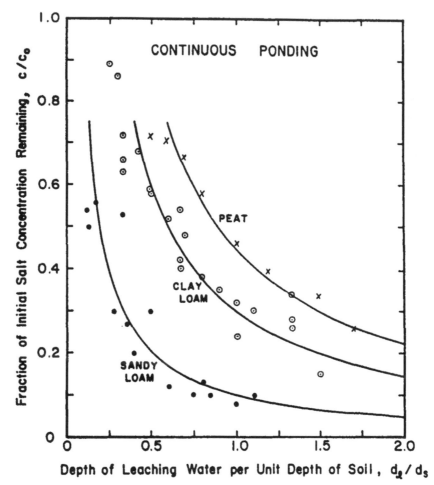

Figure 10.   Depth of leaching water per unit depth of soil required to reclaim a saline soil by continuous ponding (Rhoades, 1980).

3.   The equivalent amount of the amendment that should supply the necessary Ca^{++} to replace the excess exchangeable Na.

4.   Assuming that the applied Ca would quantitatively replace the exchangeable Na, thus, mole of Na requires half a mole of Ca. Thus, the amount of gypsum required to supply half a mole of Ca per 1 kg soil for the upper 15 cm soil depth will be:

$$\frac{\text{Molecular weight of gypsum}}{200} = \frac{172}{200} = 0.86 \text{ g/kg soil}$$
$$= 86 \times 10^{-5} \times 2.24 \times 10^{6} \text{ kg/ha}$$
$$= 1926 \text{ kg/ha or } 1.96 \text{ Mg/ha}$$

Quantities of other amendments can be calculated using the relative values in Table 14.

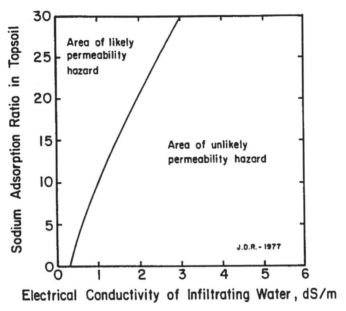

Figure 11. Threshold values of sodium adsorption ratio of topsoil and electrical conductivity of infiltrating water for maintenance of soil permeability (Rhoades).

The Soil Exchangeable Na

Because of difficulties encountered in determining a precise value of the soil exchangeable Na by the classical analytical methods, a method using ^{22}Na for this determination at 1:5 soil water ratio is simple, easy and relatively accurate (Balba et al.[118]). In Rhoades'[103] opinion, the soil exchangeable sodium can better be determined by calculating R_{Na}(SAR) in the soil solution instead of direct determination.

The U.S. Sal. Lab. staff[26] derived a mathematical relationship between the concentration of Na and Ca in the solution phase and their exchangeable fractions depending on Gapon's equation:

$$[V_{Ca}/Na]_{soln} \cdot [Na/CEC-N] = K$$

$$\text{or } [Na/V_{Ca}]_{soln} = K [Na/CEC-Na]_{exch}$$

where CEC = cation exchange capacity. The U.S. Sal. Lab. equation depended on assuming that

a) The cations present in the soil system are mainly Ca, Mg, and Na.
b) The Mg cations react similarly to Ca, i.e., both have approximately the same tightness of retention to, or ease of release from, the colloidal surface. Hence, the amounts of Ca and Mg are summed together. The modified relationship thus appears as follows:

Table 14.   Amounts of sodic soils amendments equivalent to 1 kg of pure gypsum, kg.

Amendment	Amount Equivalent to Gypsum	Remarks
Gypsum	1.00	low cost; slow acting; low electrolyte concentration
Sulfur	0.19	produces H_2SO_4 by bacteria effective for sodic calc. soils; produces gypsum *in situ;* costly
Calcium carbonate	0.58	more suitable for $CaCO_3$; poor soils
Calcium chloride dihydrate	0.85	speeds reclamation; expensive
Ferrous sulfate	1.61	similar to H_2SO_4; costly
Aluminum sulfate	1.29	

Source: Adapted from Rhoades.

$$\left[\frac{Na}{(Ca+Mg)/2}\right]_{soln} = \left[\frac{Na}{CEC-Na}\right]_{exch}$$

The left side was termed sodium adsorption ratio (SAR).

From the relation between SAR and the right side of Equation 2, the following empirical equation was arrived at relating SAR to the exchangeable Na percentage ESP:

$$ESP = \frac{100 \; (-0.0126 + 0.01475 \; SAR)}{1 + (-0.0126 + 0.01475 \; SAR)} \tag{26}$$

The U.S. Sal. Lab.[26] staff have shown that the results obtained by applying Equation 26 agreed satisfactorily with the chemically determined results.

Bower[3] modified Equation 26 to:

$$ESP = 100 \, (Na_{exch}) \, / \, (Ca_{exch} + Mg_{exch}) = KG \; SAR$$

where KG is the exchange constant called Gapon's constant.

Several investigators (Oster and Sposito[32]; Bower[3]; U.S. Sal. Lab.[26]) considered that E_{Na} (exchangeable Na ratio; Exch N/CEC-Exch Na), in $mole_c \; Kg^{-1}$ soil is in equilibrium with R_{Na}, $[Na/(Ca+Mg)/2]$ of soil saturation extract. This equilibrium was described by the linear equation:

$$E_{Na} = a \; R_{Na} + b \tag{27}$$

According to Rhoades,[103] the regression equation correction for high salinity has become:

$$E_{Na} = 0.014 \; R_{Na} - 0.058 \tag{28}$$

Schoonover's[119] method depends on shaking 5 g soil with 100 ml of gypsum solution. Thus, it is expected that under such diluted salyts conditions, Ca (Ca+Mg) replaces all exchangeable Na. Evidently, the Na_2CO_3 which might be present in the soil will also react with the applied Ca. Thus, the decrease of Ca in the gypsum solution after equilibrium is due to replacing exchangeable Na and reacting with soluble carbonates.

In order to save in the applied gypsum, it was suggested to leach the soil sample with alcohol to remove soluble carbonates before carrying out the GR evaluation. Under field conditions, an irrigation prior to the gypsum—or other amendments—would remove the soluble carbonate (Abrol and Dahiya[120]). However, in practice water infiltration in nonsaline sodic soils without amendment application is too slow.

According to Rhoades[103] and El-Gabaly,[9] an extra amount of Ca is required over and above the equivalent amount of exchangeable sodium to be replaced. When using gypsum in reclaiming the soil, the extra gypsum amount is 1/4 that of the equivalent amount of exchangeable sodium to be replaced. In other words the amount of added gypsum is increased 1.25 times. Doering and Willis[121] found that an extra amount of the amendment should be increased when high electrolytes are used in the reclamation process. This increase in GR may be the explanation of Balba et al.[118] results which showed that the exchangeable Na determined by ^{22}Na was approximately equal to the GR of the soil instead of being greater.

Prather et al.[122] suggested a dual-amendment concept to integrate the beneficial effects of more than one amendment. They applied gypsum equivalent to 75 or 80% of the excess exchangeable Na in the top 30 cm of the soil. The gypsum is mixed in that depth. The remainder of the required amendment is applied as calcium chloride pellets (in solution) or concentrated sulfuric acid followed immediately by leaching. The last two amendments improve the soil permeability and thus speed the reclamation process. The applied gypsum is dissolved later after passing the concentrated solution and supplies the major part of Ca needed for replacing the remaining exchangeable Na.

Factors Affecting the Effectiveness of Solid Amendments

The effectiveness of the solid amendments depends on their fineness, methods of application, solubility, the quantity of the applied water, the efficiency of the drainage system, and the time of their application.

Fineness

It has frequently been stated that 80–85% of the gypsum applied to the soil should pass through a 1 mm sieve (El-Gabaly[9]). El-Gibaly[123] showed no significant difference in the removed soil exchangeable Na when the sodic soil was leached with water mixed with gypsum passed through 100, 150, or 200 mesh sieves. These treatments, however, removed higher amounts of Na than with a treatment using gypsum that passed through 60 mesh sieve. Also, Chawla and Abrol[124] concluded that gypsum which passed through a 2 mm sieve and with wide particle size distribution gave better results in soils containing Na_2CO_3 to avoid formation of $CaCO_3$ coatings on the gypsum surface (Keren and Kauschansky[125]; Keren and Shainberg[126]). Rhoades[103] advised using material passed through a 6 mm sieve.

Methods of Amendments Application

Solid amendments are commonly broadcast followed by mixing with the surface soil layer by plowing or disking. Schoonover et al.[117] found that gypsum mixed with the surface 15 cm was more effective in removing the exchangeable Na than gypsum applied on the surface. In nonsaline sodic soils, which usually contain soluble carbonate, part of the applied gypsum will be consumed when reacting with the carbonate. Thus, it is not recommended to mix the amendments deep in the soil (Khosla et al.[127]).

The amendments may also be applied with the applied water. This practice is more common in applying the sulfuric acid. In Armenia as stated by El-Gabaly,[9] sulfuric acid is mixed with water at the rate of 30 M g per ha of 80% concentration to reclaim a depth of 100 cm with the addition of 1500 m^3/ha of leaching water. Solid amendments, especially gypsum, are also mixed with the applied water, especially when the water is rich in Na (Ayers and Westcot[6]).

For applying the concentrated sulfuric acid, special equipment is available to spray it on the soil surface.

Gypsum Solubility and Amount of Water

Because of the gypsum low solubility in water—0.25%—it is thought that the amount of water that should be applied is, at least, the amount necessary to dissolve the applied gypsum. Quirk and Schofield[128] and others, stated by Rhoades,[103] considered that the leachate of soil amended by gypsum applied into the topsoil was between one third- and one half-saturated with gypsum. Rhoades[103] thus recommended dividing the amount of water and gypsum into portions to achieve the soil reclamation over a period of 2 to 5 years.

Efficiency of Drainage

It should be pointed out that the solubility of gypsum when incorporated in the sodic soil is greatly increased as the dissolved Ca^{++} replaces exchangeable Na. The latter forms Na_2SO_4 and is removed from the soil-gypsum system by drainage. Thus, it is not the solubility of gypsum in water that controls its solubility in the sodic soil system. Hira et al.[129] and Jury et al.[111] developed equations for predicting gypsum dissolution and water requirement to achieve reduction in exchangeable sodium.

It has been stated above that leaching should follow the application of the amendments to the sodic soils. Leaching signifies the presence of efficient drainage. Also, it has been stated above that drainage is necessary to remove the sodium salt produced from the exchange reaction, which in the case of gypsum is Na_2SO_4.

Time of Gypsum Application to Saline Sodic Soils

The time of gypsum addition to the saline sodic soils has also been a matter of argument. Several soil scientists used to recommend its application after the soil was partially leached. Balba and El-Laithy[130] showed that the removal of salts by leaching was faster in soil columns mixed with gypsum than those without gypsum (Abrol et al.[31] reached the same conclusion, Figure 12). Also, application of gypsum before leaching effected replacement of exchangeable Na. They pointed out that it is known that the saline sodic soils might turn to sodic soils if their soluble salts are leached out. If this change to sodicity takes place, reclamation of the soil might be retarded. Verhoeven[131] reached the same conclusion under the Dutch soil conditions.

Cropping During Leaching

In virgin saline areas, salinity may be too high to permit crop growth. Under such conditions, cropping has to wait until salinity is decreased to a level which permits salt-tolerant crops to grow. One of the successful winter crops is the Italian ryegrass. In summer, *Panicum crus* (Barnyard grass) and *Panicum repens* may be grown as fodders. Rice is also successful. These crops require considerable amounts of water. Thus, leaching takes place during their growth. The salt level after these crops is usually greatly decreased allowing less tolerant crops to grow.

Salinity tests should be carried out at the end of the crop season. The selected crop should tolerate the salinity level of the soil.

Cropping during leaching improves the efficiency of the leaching process because the roots penetrate the soil and enhance water infiltration. Plowing in this stage should be deep. Also, subsoiling is recommended.

Post Reclamation Practices

Tillage

An important goal of salt-affected soils management is not to allow the soil to be resalinized. Soil resalinization may take place as a result of groundwater rise and its subsequent evaporation, leaving its salt content on the soil surface. Such mechanism may be avoided by hindering water capillary rise and enhancing water infiltration through the soil. Several technologies may be recommended for this purpose:

1. Plowing or subsoiling improves deep water penetration. However, after one or two irrigations the soil surface reverts to its original condition, though sufficient water infiltrates and may improve the stored water and the crop yield, Ayers and Westcot.[6]
2. Deep tillage may be carried out prior to planting. Ayers and Westcot[6] recommended that this operation be done when the soil is dry enough to shatter and crack and not when the soil is wet.
3. Cultivation is a common practice for weed control. Aeration also takes place as a result of cultivation which is reflected in improving plant growth.

In soils suffering from water infiltration problems, cultivation or deep tillage is practiced. Both operations roughen the soil surface, thus the water surface flow slows down allowing more time for infiltration. This effect of cultivation usually lasts for a period of one or two irrigations after which another cultivation may be needed (Ayers and Westcot[6]). Also, cultivation breaks the crust formed on the soil surface which further improves infiltration.

A practice recommended when the irrigation water has low salinity is to cultivate the soil before each irrigation. This practice roughens the soil and opens cracks and air spaces which improves infiltration (Ayers and Westcot[6]).

1. The effectiveness of this practice depends on the depth to water table, pore size distribution, climatic conditions, crop cover, rate and method of residues application, and mulching or mixing with the soil.
2. Gravel placement on the soil surface best suits the orchards since the gravel layer creates a problem when cultivating the soil.
3. Surface tillage or cultivation may also be practiced to control upward soil water movement to the soil surface. Massoud[132] recommended that tillage be carried out at an early stage while the evaporation rate is high and not after considerable surface drying had taken place which reduces the expected beneficial effect of tillage.
4. The soil conditioners are chemicals applied to the soil to improve its physical characteristics. Among these conditioners are hydrolyzed starch, polyacrylonitrile graft copolymer (H-Span or Super Sluper), venyl alcohol-acrylic acid copolymers and polyacrylamides, and others. They are polymers synthesized to be nonionic, anionic, or cationic. When a polymer is applied to the soil it is usually adsorbed on the surface of the soil single particles. This adsorption alters the particles, relation with water and ions in the surrounding solution. Attachment of one polymer molecule to several soil particles may also take place. Thus, the formation of interparticle bonds enhances the flocculation of a dispersed system or stabilizes existing unstable or metastable arrangements of particles.

When the conditioner is used in clay or sodic soils according to established procedures the following improvements may take place:

*   Increase pore spaces in soils.
*   Increase water infiltration into soils.
*   Make the soil friable easy to cultivate.
*   Make the soil dry quicker after rain or irrigation, so that the soil can be worked sooner.

These improvements result in earlier seed emergence, easier weed removal, less plant diseases related to poor soil aeration, and decreased energy required for tillage (Wallace and Nelson[133]; Wallace and Abuzamzam[134]; Wallace and Wallace[135]; Wallace et al.[136]).

Cropping

Crop Selection

To utilize reclaimed salt-affected soils, selection of crops is an effective means at least partially to avoid resalinization.

It has been stated above that plants differ considerably in tolerating salinity, sodicity, or waterlogging conditions. The U.S. Sal. Lab.[26] published lists of crops and the salinity levels that might cause 50% decrement in their yield. In this regard, Rhoades[103] stated that the data concerning crops salt tolerance published by the U.S. Sal. Lab.[26] do not represent salt tolerance for germination and early seedling growth. He concluded that salt-tolerance data should be taken as a relative rating of the tolerance of crops to salinity. Maas and

Hoffman[52] summarized salt tolerance of crops from various sources. Research work in breeding and genetic engineering enhances hopes to have cultivars of economical crops that can tolerate high salinity. Saline soils which cannot be reclaimed due to variable circumstances may be left for grazing pastures. Malcolm et al.[137] showed that *Atriplex rhagodioides, A. undulata,* and *Mavieana brevifolia* are well suited to forage production on saline soils in southwest Australia.

In addition, the following practices are recommended:

- Crops which require flooding such as paddy rice help leach salts from the soil. The soil is usually less saline after harvesting these crops than before their planting. On the contrary, crops which require a long hot dry period for maturity such as barley enhance resalinization, especially under shallow groundwater table. Crops which require frequent irrigation do not allow resalinization. An extra amount of water—the leaching requirement—is usually applied to prevent salt accumulation in the root zone. Table 16, (Rhoades[103]) gives the leaching fraction under variable salinity levels to obtain yields without considerable reduction and for various crops.
- Increasing plant density is usually recommended to overcome the salt adverse effect on seed germination (Hamdy[138]).
- Because Cl-toxicity is common due to the dominance of Cl-salts in the soils and waters, Cl-tolerant crops are usually preferred. Application of phosphorus fertilizers is recommended to counteract the Cl effect. The same goes for B-toxicity if B salts are present in problem amounts.

Seed Placement

Because the germination stage is critical for many crops as they are more sensitive to salinity in this stage, the seeds should be placed where the soil salinity is at its minimum. Thus, chances for the seeds to emerge are improved which is reflected on the crop stand and subsequently on the yield. In this regard, plant population should be increased when seeded in saline soils (Hamdy[138]).

The flood and basin methods of irrigation leach salts from the soil surface; under furrow irrigation, salts accumulate on the furrows' tops. Bernstein et al.[139] recommended that seeds should not be placed in the upper center of the furrow where salts are concentrated. They recommended placing the seeds in double-row beds, each row is located at the furrow shoulder away from the area of greatest salt accumulation. These are other seed placement alternatives schematically shown in Figure 13.

Irrigation[*]

Preplanting Irrigation

A heavy preplanting irrigation to leach salts that might have accumulated at the soil surface is recommended (Ayers and Westcot[6]). Stevens stated, after describing cotton cultivation on salt-affected soils, that it was most likely that the first irrigation 4 to 8 days before planting was the most important factor determining the decrease in EC and pH. The soil content of Cl⁻ was almost washed out of the root zone of the fine-textured soil.

---

[*]   See Chapter 5.

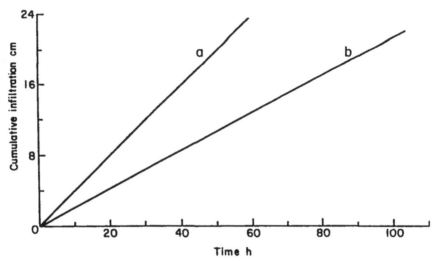

Figure 12. Effect of gypsum application on cumulative infiltration in a saline soil. a - with, b - without gypsum (Abrol et al.).

Decrease of SO_4^- had also taken place. Several studies have shown that the first irrigation before planting was the most important factor determining the decrease in EC.

Amount of Water

To prevent the excessive accumulation of salts as a result of irrigation, especially with waters of poor quality, an additional increment of water over and above that required to meet evapotranspiration must be passed through the root zone. This additional water increment is termed the leaching requirement.

Irrigation Frequency

More frequent irrigations are usually recommended for crops grown in reclaimed saline soils than for crops grown under nonsaline conditions:

a.   Plants usually absorb water at a much higher rate than cations and anions. There-fore, salts accumulate in the root zone. Water is applied to restore the salt concen-tration around the roots to its original level.

b.   The salt concentration in the soil solution also increases as the soil progressively dries out by evapotranspiration. Thus, its osmotic pressure increases.

c.   Keeping the soil at a higher soil moisture content by frequent irrigations prevents the concentration of salts in the soil solution and minimizes their adverse effects.

Methods of Irrigation

Sprinkler irrigation has proved more efficient than flood irrigation in leaching-soluble salts as small amounts of water can be applied more frequently (Nielsen et al. [102]).

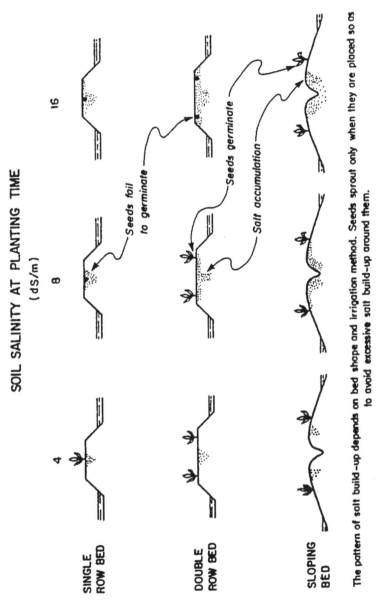

Figure 13.   Bed shapes and salinity effects (from Ayers and Westcot).

Because water is supplied continuously by drip method in the immediate vicinity of plant roots, the soil moisture is kept continuously high in the root zone. A comparison was carried out by Goldberg et al.[140] between the effect of irrigation using the sprinkler and the drip methods in a case of irrigation with two waters having different salinity levels.

Drainage*

As stated above, an efficient drainage system is essential not only for reclamation but also for preventing redeterioration of the soil. The drainage system may be a network of open ditches or tiles made of earthware, cement, or plastic. Tube wells can also be used for drainage. An outlet for the drainage water should be provided.

Fertility Management of Salt-Affected Soils

The effects of salinity and sodicity on nutrients' availability to plants and on the ability of plant roots to absorb nutrients were discussed in a previous part of this book. Because of the low productivity problem in the salt-affected soils, fertilizers are applied to counteract the conditions which limit the plant absorption of nutrients.

In reviewing studies on fertility-salinity interactions, two main approaches are usually encountered: the physiological and the agronomical approaches. The physiological studies are conducted to find out basic facts which might explain the adverse effects on the plant biological processes and on the absorption of nutrient elements in the presence of known kinds and concentrations of salts. Evidently such studies constitute the basic relationships which govern plant responses to fertility and salinity levels. On the other hand, the agronomic studies are concerned with the response of a crop under specified soil salinity conditions to increments of fertilizers. They are mainly concerned with counteracting the adverse effect of the soil salinity, sodicity, or waterlogging on the production of a particular crop on a specified soil. They might go a little further to calculate the optimum rate of fertilizer application for a specified situation of crop, soil, and cost of production. These economical aspects may be intensified to calculate the monetary return per hectare, or per dollar spent on fertilization.

The presence of excess soluble salt concentrations, high sodicity, poor aeration, and permeability problems, which prevail in saline, sodic, or waterlogged soils, impose several constraints on the productivity of these soils. Because numerous factors are involved in plant response to fertilizers under saline, sodic, or waterlogged conditions, several investigations were conducted to reach a sound fertility management for these soils but contradicting results were frequently encountered. Bernstein et al.[139] stated that one has to compare the benefits which might be gained from removing the excess salts from the saline soil to improve its productivity with the benefits gained by applying fertilizers to the saline soil.

Soil tests for nutrients under saline conditions correlate well with values under nonsaline conditions. However, efficiencies of fertilizers applied to salt-affected soils are lower than when applied to nonsaline soils (Amer et al.[141]; Balba and Soliman[142]).

A decrease in the ability of the plants to absorb K or NH_4 usually takes place in saline soils containing excess Na, Mg, or Ca. Also, P absorption may be decreased in presence

---

*   See Chapter 5.

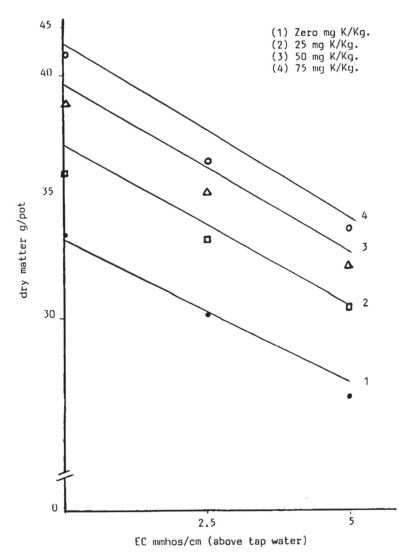

Figure 14. Effect of salinity on dry matter yield of tomato under K increments (Balba et al.,[84] 1982).

of excess Cl⁻ or SO_4^- as stated above. Application of K, NH_4, or P fertilizers not only corrects their deficiencies but also decreases the adverse effects on the plants of Na, Cl or SO_4^- (Figure 14).

It has been stated above that forms of nutrients may be changed in salt-affected soils. Processes related to the nutrient formation in the soil may also be affected. McClung and Frankenberger[143] showed that a reduction in the nitrification rate and NO_2^- accumulation was detected in a soil exposed to salinity stress after being treated with ammonium fertilizer. Also, Jurinak and Wagenet[144] showed that salinity had affected N fixation by *Rhizobium* spp. bacteria. An explanation for this effect may be:

- The high osmotic potential in saline soils may cause water loss from the nodules to the soil solution, Jurinak and Wagenet.[144]
- High salt level impairs the metabolic activity in the nodule cortical cells, Sprent.[145]

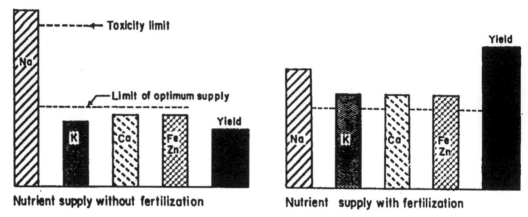

Figure 15. Schematic diagram showing the effect of fertilization for correction of an unbalanced and insufficient nutrient supply to plants in saline soils; the columns indicate the plant contents (Finck,[147] 1977).

The organic manures have two functions. They improve the soil physical properties, especially in reclaimed nonsaline sodic soils and supply the plants with nutrients.

A schematic drawing showing the possible effect of fertilization on plants growing on saline soils is shown in Figure 15.

The Salt Index of Fertilizers

The more soluble the fertilizers are, the higher their osmotic potential which develops in the root zone. In other words, soluble fertilizers increase the soil salinity in the root zone. Therefore, it is recommended not to apply the fertilizers in direct contact with the seeds or the seedlings roots. Ayers and Westcot[6] give a salt index for various fertilizers. This salt index expresses the potential salinity effect of fertilizers relative to sodium nitrate as follows:

$$\text{The salt index (SI)} = P/P_1 \times 100$$

where P is the osmotic potential resulting from any fertilizer and P_1 is the osmotic pressure created by the addition of $NaNO_3$.

The salt index of highly soluble fertilizers such as RCl or NH_4NO_3 is high being 114 and 104, respectively. Phosphate fertilizers which have low solubility have a low SI of 8 to 10 (Table 15).

Lunin and Gallatin[148] added N, P, or K fertilizers each at the rate of 168 kg/ha to a soil salinized to EC_e 1.7, 3.5, 5.5, or 7.8 dS m^{-1}. They showed that N fertilizer had considerably increased EC_e while the soil salinity remained approximately constant in the case of K application and tended to decrease in the case of P application, especially with the low levels of EC_e (Table 16).

Table 15.   Relative effect of fertilizer materials on the soil solution.

Material	Salt Index[1]	Partial Salt Index per Unit of Plant Nutrient
Anhydrous ammonia	47.1	0.572
Ammonium nitrate	104.7	2.990
Ammonium nitrate-lime	61.1	2.982
Ammonium phosphate (11-48)	26.9	2.442
Ammonium sulphate	69.0	3.253
Calcium carbonate (limestone)	4.7	0.083
Calcium cyanamide	31.0	1.476
Calcium nitrate	52.5	4.409
Calcium sulphate (gypsum)	8.1	0.247
Diammonium phosphate	29.9	1.614
Dolomite (calcium and magnesium carbonates)	0.8	0.042
Kainit, 13.5%	105.9	8.475
Kainit, 17.5%	109.4	6.253
Manure salts, 20%	112.7	5.636
Manure salts, 30%	91.9	3.067
Monoammonium phosphate	34.2	2.453
Monocalcium phosphate	15.4	0.274
Nitrate of soda	100.0	6.060
Nitrogen solution 37%	77.8	2.104
Nitrogen solution 40%	70.4	1.724
Potassium chloride, 50%	109.4	2.189
Potassium chloride, 60%	116.3	1.936
Potassium chloride, 63%	114.3	1.812
Potassium nitrate	73.6	5.336
Potassium sulphate	46.1	0.853
Sodium chloride	153.8	2.899
Sulphate of potash-magnesia	43.2	1.971
Superphosphate, 16%	7.8	0.487
Superphosphate, 20%	7.8	0.390
Superphosphate, 45%	10.1	0.224
Superphosphate, 48%	10.1	0.210
Uramon	66.4	1.579
Urea	75.4	1.618

[1]   The salt index is for various fertilizer materials when applied at equal weights. Sodium nitrate, with a salt index of 100, is used as a base for the index.

Source: Data taken from Rader (1943).

Table 16.   Effect of added fertilizers and salinity on selected soil parameters.

Sal. Level	None	N	P	K	NP	NK	Pk	NPK
	Ec, dS m^{-1}							
A	1.7	2.0	1.8	1.3	1.3	2.4	1.4	1.5
B	3.5	4.3	3.7	3.8	3.8	5.1	3.8	4.2
C	5.5	6.4	5.6	6.8	6.8	7.0	6.5	7.2
D	7.8	9.2	8.0	8.4	8.4	10.3	7.9	8.8

Source: Adapted from Lunin and Gallatin.[132]

Table 17.   Effect of saturation with exchangeable Na, Mg, or Ca on soil physical properties.

	Na	Mg	Ca
Ex. Ca, meq/100 g soil	1.5	2.64	30.98
Ex. Mg, meq/100 g soil	1.61	27.67	1.02
Ex. Na, meq/100 g soil	30.0	0.50	0.72
Ex. K, meq/100 g soil	1.4	0.86	0.92
M.e. %	49.12	36.80	37.03
P.W.P. %	25.86	13.03	12.74
S.F. %	27.49	86.23	86.23
D.C. %	72.51	13.80	14.00

M.E. = Moisture Equivalent; P.W.P. = Permanent Wilting Point; S.F. = Structure Factor; D.C. = Dispersion Coefficient.
Source: El-Ashtar.[156]

Magnesium Soils

The properties of Mg soils have been a controversial matter since the early 1930s. Several soil scientists showed that soils containing high exchangeable Mg percentages acquire the characteristics of the sodic soils.

Somewhat later, Sherman[149] and Klages[150] stated that magnesium and sodium are responsible for the dispersion, stickiness, and plasticity of the soils. Yadav and Girdhar[151] reported more clay particles dispersion and low hydraulic conductivity when Mg/Ca ratio increases at specified SAR and salt concentration of the leaching water. On the other hand, according to Kelley,[152] Martin[153] as early as 1929 showed that the highly productive Yolo soil of the University Farm at Davis, California, USA, contains as much exchangeable Mg as exchangeable Ca. Several other investigators (Reeve et al.[154]; Brooks et al.[155]) had shown that Mg effect on the soil physical properties, especially the stability of structure, is more similar to calcium effect than to sodium effect. In Egypt, the soils of the northern part of the Nile Delta are characterized by fine impermeable clay, rich in soluble and exchangeable Mg. Soil researchers considered these properties as a result of their exchangeable Mg and suggested the application of gypsum to improve these soils. El-Ashtar[156] saturated soil samples by Na, Mg, or Ca and determined several physical properties which showed that the Mg-saturated soil properties are similar to those of the Ca-saturated soil (Table 17).

It is believed that the North Delta soils contain more exchangeable Mg than soils of the South Delta because soils of the North Delta region are formed in the presence of sea and lake water, rich in soluble Mg. The deposition process at the farthest deposition region, close to the discharge of the Nile water into the Mediterranean Sea is characterized by its slowness due to the finer particles of the deposits and the deposition process in the presence of water. These conditions result in soil with low hydraulic conductivity. It may also be stated that the dominant clay mineral of the region is montmorillonitic which swells upon moistening. Thus, water infiltration is further impeded. However, the Mg soils controversy is still not settled in various regions.

REFERENCES

1. Kovda, V., Alkaline soda-saline soils. Sodic Soils Symposium, Tal. es Agrok. Vol. 14, 15, 1965.
2. Rhoades, J.D., Soil Salinity, Causes and Control. Chapter 4 in Techniques for Desert Reclamation. ed. A.S. Goudie, John Wiley & Sons, New York, 1990.
3. Bower, C.A., Cation exchange equilibrium in soils affected by sodium salts. Soil Sci., 88:32, 1959.
4. Doering, E.J. and F.M. Sandoval, Hydrology of saline seeps in the Northern Great Plains. Trans. A.S.E.A. 19:856, 1976.
5. Balba, A.M. and M.F. Soliman, Salinization of homogeneous and layered soil columns due to upward movement of saline ground water and its evaporation. Agrochimica, 13:542, 1969.
6. Ayers, R. and D.W. Westcot, Water Quality for Irrigation. FAO Irrig. and Drainage Paper, No. 29 Rev. 1 p. 173, 1985.
7. Balba, A.M. and M.F. Soliman, Distribution of salts in sand due to capillarity, evaporation transpiration and diffusion. Alex. J. Agric. Res., 26:247, 1978.
8. Eaton, F., Significance of Carbonates in Irrigation Waters. Soil Sci., 69, 123, 1950.
9. El-Gabaly, M.M., Reclamation and management of salt affected soils. In salinity Seminar, Baghdad, Irrigation and Drainage Paper No. 7, FAO, Rome, 1971.
10. Whitting, L.D. and P. Janitzky, Mechanism of formation of sodium carbonate in soil. J. Soil Sci., 14:322, 1963.
11. Beek, C.G. and N. Van Breemen, The alkalinity of alkali soils. J. Soil Sci., 24:129, 1973.
12. Kelley, W.P., Alkali Soils, Their Formation, Properties and Reclamation, Reinhold Pub., New York, p. 68, 1951.
13. Kovda, V., To combat salinization editorial Climatic Change, 2:103, 1979.
14. Dregne, H., UN Conf. Desertification 74/3 Add. 2, 1977.
15. FAO/UNESCO Desertification Map of the World, A- Conf. 72-2, 1974.
16. Balba, A.M., A Working Report on Desertification. Minimum Management Programme to Combat World Desertification. UNEP Consultancy, 1980. Alex. Sci. Exch. Vol. 1, No. 2, 41, 1980 and Vol. 2, No. 1, 23, 1981.
17. FAO, Water for Agriculture, UN-Water Conf., Mardel Plat (W/K330), 1977.
18. Massoud, F.I., Salinity and alkalinity as soil degradation hazards. FAO/UNEP AGLSD/74/10, Rome, 1974.
19. Ehlrich, W. and R.E. Smith, Halomorphism of some clay soils of Manitoba, Canada. Can. J. Soil Sci., 38:103, 1958.
20. Blazhni, E.S., Cited in Bernstein,[20] 1962.
21. Zavaleta, A.G., The Peruvian coastal zone. Int. Symp. on Sodic Soils. Tal. es Agrok., 14, 415, 1965.
22. Agaev, B.M. and G.S. Kulieva, Salt liberation capacity of sodic soils of the Karabakh plain, Azerbaijan. Int. Symp. on Soda Saline Soils, Yerevan Tal. es Agrok., 18, 345, 1968.
23. Borovski, Solonetz and Sodic soils in Kazakhstan. Sodic Soils Symp. Tal. es Aqrok., 14, 245, 1965.
24. Szabales, I., Degradation of irrigated soils in Hungary. 7th Int. Soil Sci. Cong. Vol. 1, 638, 1960.
25. Ayers, A.D., Saline and Sodic Soils of Spain. Soil Sci., 90:133, 1960.
26. U.S. Sal. Lab. (Richards, L.A., ed.), Diagnosis and Improvement of Saline and Alkali Soils. USDA. Hand Book No. 60, p. 160, 1954.
27. Rhoades, J.D. and R.D. Ingvalson, Determining salinity in field soils with soil resistance measurements. Soil Sci. Soc. Amer. Proc., 35:54, 1971.

28. Rhoades, J.D., Measuring and monitoring soil salinity. Paper No. 45 in Water, Soil and Crop Management Relating to the Use of Saline Water. FAO, AGL/MISC/16/90, Rome, pp. 71, 1990.

29. Rhoades, J.D., N.A. Manteghi, P.J. Shouse, and W.J. Alves, Estimating soil salinity from saturated soil paste electrical conductivity. Soil Sci. Soc. Amer. 53:433, 1989.

30. Rhoades, J.D., Soil salinity assessment, Recent advances and findings. U.S. Sal. Lab., 1993, Private communications.

31. Abrol, I.P., J.S.P. Yadav, and F.I. Massoud, Salt-Affected Soils and their Management. FAO - Soil Bull. 39, FAO, Rome, 1988.

32. Oster, J.P. and G. Sposito, The Gapon coefficient and the exchangeable sodium percent-age, sodium adsorption ratio relation. Soil Sci. Soc. Amer. J., 44:258, 1960.

33. Rao, T.S., A.L. Page, and N.T. Coleman, The influence of ionic strength and ion-pair formation between alkaline earth metals and sulfate on Na-divalent cation exchange equilibrium. Soil Sci. Soc. Amer. Proc., 32:639, 1968.

34. Balba, A.M. and A.H. Balba, Sodium selectivity of bentonite clay in presence of chloride and sulfate expressed by thermodynamic parameters. Egyptian J. Soil Sci., 16: 179,1973.

35. Balba, A.M. and A.H. Balba, Effect of anion composition of the solution phase on the cation exchange reaction. Int. Symp. Salt-Affected Soils Tranc., p. 361, Cairo, 1972.

36. Sposito, G. and S.V. Mattigod, On the chemical foundation of the sodium adsorption ratio. Soil Sci. Soc. Amer. J., 41:323, 1977.

37. Nakayama, F.S., Evaluation of the sodium-calcium exchange constants in chloride and sulfate soil systems, the associated and nonassociated models. Soil Sci., 119:410, 1975.

38. Suarez, D.L., Relation between pHc and SAR and an alternate method of estimating SAR of soil drainage waters. SSSAJ, 45:469, 1981.

39. FAO, Prognosis of Salinity and Alkalinity. FAO Soils Bull. 31, FAO, Rome, 19, 1976.

40. Balba, A.M., Predicting soil salinization, alkalization and waterlogging. FAO Soils Bull. 31, p. 241, 1976.

41. Rhoades, J.D., Measuring, mapping and monitoring field salinity and water table depths with soil resistance in Prognosis of Salinity and Alkalinity. FAO Soils Bull. 31, p. 159, 1976.

42. Rhoades, J.D. and S.D. Merril, Assessing the suitability of water for irrigation, Theoret-ical and empirical approaches. FAO Soils Bull. 31, p. 69, 1976.

43. US Soil Survey Staff, Keys to Soil Taxonomy SMSS-Tech. Monograph No. 6, Chapters 4–7, 1987.

44. Walter, H., The adaptation of plants to saline soils. Tehran Symp., UNESCO Pub., Paris, p. 129, 1961.

45. Radin, J.W., Physiological consequences of cellular water deficit. Osmotic adjustment. Chapter 6 B in Limitations to Efficient Water Use in Crop Production, p. 267, 1983.

46. Fiscus, E.L., Water transport and balance within the plant: Resistance to water flow in roots. Chapter 4 C. in Taylor et al. (Ed.).

47. Repp, G., The salt tolerance of plants. Basic research and tests. Tehran Symp., UNESCO, Pub., Paris, p. 153, 1961.

48. Magistad, O.C., A.D. Ayers, C.H. Wadliegh, and H.G. Gauch, Effect of salt concentra-tion, kind of salt and climate on plant growth in sand culture. Plant Phys. 18, 151, 1943.

49. Murthy, K.S. and K.V. Janardhan, Physiological consideration for selection and breeding of varieties for saline and alkaline tracts. Oryza J., 8:85, 1971. In Abrol et al., 1971.

50. Bernstein, L., Salt-Affected soils and plants. Paris Symp. on Problems of Arid Zones. UNESCO Pub., Paris, p. 139, 1962.

51. University of Calif. Com. of Consultants, Cited in Ayers and Westcot,[19] 1985.

52. Maas, E.V. and G.J. Hoffman, Crop salt tolerance current assessments. U.S. Sal. Lab. Pub. 1976. Cited in Ayers and Westcot,[19] FAO Irrig. and Drainage Paper Nov. 29, Rev., 1985, p. 36.

53. Pearson, G.A., A.D. Ayers, and D.L. Eberhard, Relative salt tolerance of rice during germination and early seedling development. Soil Sci., 103:151, 1966.
54. Pearson, G.A., The salt tolerance of rice. Int. Comm. Newsletter, 10:1, 1961.
55. Kaddah, M.T. and S.I. Fakhry, Tolerance of Egyptian rice to salt. Soil Sci., 91:113, 1961.
56. Cerda, A.M., F.G. Fernandez, and M.G. Guillen, Foliar content of sodium and chloride on citrus rootstocks irrigated with saline waters. Proc. Int. Conf. on Managing Saline Water for Irrigation, Texas Tech. Univ., Lubbock, Texas, p. 155, 1977.
57. Bernstein, L., Salt Tolerance of Fruit Crops. USDA Agric. Inf. Bull. 292, p. 8, 1965.
58. Gauch, H.G. and C.H. Wadliegh, The salt tolerance and chemical composition of Rhodes and Dalis grasses in sand cultures. Bot. Gazette, 112:259, 1951.
59. Shukla, U.C. and A.K. Makhi, The ameliorative role of zinc on the growth of maize (Z. maus L.) under salt-affected soil conditions. Int. Symp. Salt Affected Soil Condition Proc. 362, 1980.
60. Devitte, D., W.M. Jarrel, and K.L. Stevens, Sodium-potassium ratios in soil solution and plant response under saline conditions. Soil Sci. Soc. Am. J., 45:80, 1981.
61. El-Fakharany, Y.M., Soil Fertility in Potassium and the Effect of Saline Water on the Efficiency of Potassium Fertilization in Ismailliah Governorate. M.SC. Thesis, University of Suez Canal, 1981.
62. Balba, A.M. and M.R. Shabana, The efficiency of different nitrogen forms as tested by barley seedlings. Plant and Soil, 37:33, 1971.
63. Zhukovskaya, N.V., Uptake and accumulation of phosphate by plants in salinized soils. Soils and Fertilizers, 36, 241, 1973.
64. Torres, C. and F.I. Bingham, Salt-tolerance of Mexican wheat. I. Effect of NO_3 and NaCl on mineral nutrition, growth and grain production of four wheats, Soil Sci. Soc. Am. Proc., 37:711, 1973.
65. Bernstein, L., Osmotic adjustment of plant in saline media. II- Dynamic phase. Am. J. of Botany, 50:60, 1963.
66. Jains, B.E., Adjustment mechanisms of plants subjected to varied osmotic pressures of nutrient solutions. Soil. Sci., 101:180, 1966.
67. Bernstein, L., L.E. Francois, and R.A. Clark, Interactive effects of salinity and fertility on yields of grains and vegetables. Agron. J., 66:412, 1974.
68. Balba, A.M. and A.H. Balba, Effect of anion increments on the composition of barley seedlings. Alex. J. Agric. Res., 23(2):349, 1975.
69. Balba, A.M. and M.R. Shabana, The efficiency of different N forms as tested by barley seedlings. Soil Sci. Soc. Am. Proc., 53:598, 1971.
70. Balba, A.M. and M.F. Soliman, Real and potential evapotranspiration under different saline conditions. Alex. J. Agric. Res., 26, 739, 1978.
71. Balba, A.M. and F. El-Etriby, The quantitative expression of the effect of water salinity on plant growth and nutrient absorption. Int. Saline Soils Symp., Karnal, India, 451, 1982.
72. Myakova, A.N., The water regime and salt tolerance of millet and wheat on saline soil, 1954. Cited in Strogonov, p. 100, 1962.
73. Lunin, J., M.H. Gallatin, and A.R. Batchelder, The Effect of stage of growth at time of salinization on the growth and chemical composition of beans. I- Total salinization accomplished in one irrigation. Soil Sci., 91:194.
74. Strogonov, B.P., Physiological Basis of Salt Tolerance of Plants as Affected by Various Types of Salinity. Translated and edited by Poljakoff, A.P., Mayber and Mayer A.M., pp. 98–128, 1964.
75. Shalhavet, J. and L. Bernstein, Effect of vertically heterogenous soil salinity on plant growth and uptake. Soil Sci., 106:85, 1968.

76. Balba, A.M. and S. El-Khatib, Available soil manganese in corn plants under variable soil conditions. Alex. Sci. Exch., Vol. 1, 51, 1980.

77. Thabet, A.Y.G., Iron Relationships with soils and plants. M.Sc. Thesis, University of Alexandria, Egypt, 1976 (Unpublished).

78. Balba, A.M., An Evaluation of Some Quantitative Factors Affecting Phosphate Solubility in Calcareous Soils. M.Sc. University of Arizona, U.S., 1954 (Unpublished).

79. Chu, W.K. and S.C. Chang, Forms of phosphorus in the soils of Taiwan. J. Agric. Ass., China 30, 1, 1960, Cited in Paliwal and Maliwal,[50] 1972.

80. Sinha, S.D. and P.B. Bhattorcharya, Studies on a phosphorus-deficient soil in Singhbum District of Bihar, Symp. Fertil. Indian Soils Bul., National Inst. Sci. India, 26:37, 1962. Cited in Paliwal and Maliwal,[50] 1972.

81. Rankov, V., The salinization of the soils and the development of the nitrogen fixation microorganisms. Agrochimica 8:330, 1964. Cited in Paliwal and Maliwal,[50] 1972.

82. Bains, S.S. and M. Firemen, Growth of sorghum in two soils as affected by salinity, alkalinity, nitrogen and phosphorus levels, Ind. J. Agron., 13:103, 1988.

83. El-Shakweer, M.H. and S.A. Abdel Ghaffar, Nitrogen balance studies for some salt-affected soils cropped with cotton. Int. Salt-Affected Soils. Proc. 909, 1972.

84. Balba, A.M., S.K. Atta, and Y. El-Fakharany, Potassium relations with soil and plant under saline conditions. Alex. Sci. Exch. 1:145, 1980.

85. Chhabra, R., I.P. Abrol, and M. Singh, Leaching losses of phosphorus in sodic soils. Int. Symp. Salt Affected Soils Proc., 418, 1972.

86. Paliwal, K.W. and C.L. Maliwal, Effect of fertilizers and manure on the mineralization and availability of nitrogen to barley irrigated with different quality waters. Int. Symp. on Salt-Affected Soils Proc., 709, 1972.

87. Bernstein, L. and G. Ogata, Effect of salinity on nodulation, nitrogen fixation and growth of soybean and alfalfa. Agron. J., 58:201, 1966.

88. Woodruff, J.R., J.T. Ligon, and B.R. Smith, Water table depth interaction with Nitrogen rates on subirrigated corn. Agron. J., 76:280, 1984.

89. Uhvits, R., Effect of osmotic pressure on water absorption and germination of alfalfa seeds. Am. J. Botany, 33:278, 1946.

90. El-Gabaly, M.M., Studies on salt tolerance and specific ion effects on plant. UNESCO Symp. on Salinity Problems Proc., 169, 1961.

91. Hale, V.Q. and A. Wallace, Bicarbonate and phosphorus effects on uptake and distribution in soybean of iron chelated with ethylene diamine di-O-hydroxiphenyl acetate. Soil Sci., 89:285, 1968.

92. Rhoades, J.D. and J. Loveday, Salinity in irrigated agriculture, 1990, in Rhoades, 1990.

93. Robinson, Frank E., Periodical and actual yield decline from fifty percent increase in salinity of the Colorado. Int. Symp. Managing Saline Water for Irrig. Proc., 170, 1977.

94. Deurt, C.T., Transpiration and crop yields. Inst. Biology and Chem. Res. on Field Crops and Herbage, No. 64, Cited in Hanks, R.J., 1974.

95. Hanks, R.J., Model for predicting plant yield as influenced by water use. Agron. J., 69:660, 1974.

96. Childs, S.W. and R.J. Hanks, Model of soil salinity effects on crop growth. Soil Sci. Soc. Am. Proc. 39:617. 1975.

97. Nimah, M. and R.J. Hanks, Model for estimating soil water, plant and atmospheric interrelations. I-Description and sensitivity. Soil Sci. Soc. Am. Proc., 37:522, 1975.

98. Szbolss, I., Investigation with radioactivity tracers of sodium carbonate in soils. Int. Symp. Isotope Rad. Ankara, 37, 1955.

99. Balba, A.M., Quantitative study of the salinization of soil columns. Sodic Soils symp. Tal. Agrok. (Hungary), 14:335, 1965.

100. Gardner, W.R. and R.H. Brooks, A descriptive theory of leaching. Soil Sci., 83:295, 1957.

101. Balba, A.M. and H. Bassiuny, Calculation of soluble Na at variable depths in sand columns after leaching using radioactive tracing. J. Isot. and Rad. (Egypt), 9:71, 1977.

102. Nelsen, D.R., J.W. Biggar, and J.N. Luthin, Desalinization of soils under controlled unsaturated conditions. Int. Com. on Irrig. and Drainage, 6th Cong., New Delhi, India, 1966.

103. Rhoades, J., Reclamation and management of salt-affected soils. Proc. of the First Annual Western Provinces Conf., pp. 123, 1982.

104. Van der Molen, W.H., Desalination of saline soils as a column process. Soil Sci., 81:19, 1956.

105. Kovda, V., Optimum level of water table. FAO Seminar on Waterlogging in Relation to Irrigation and Salinity Problems. Lahore, Pakistan, p. 122.

106. Bryssine, Cited in Durand, S.H., The quantity of irrigation water. Sols Africans, 4:53, 1959.

107. Pearson, G.A. and A.D. Ayers, Production Res. Report No. 43, USDA Sal. Lab., 1960.

108. Bresler, E., B.L. McNeal, and D.L. Carter, Saline and Sodic Soils. Principles Dynamics Modeling. Springer-Verlag, New York, in Rhoades,[103] 1982.

109. Oster, J.D. and A.D. Halverson, Saline seep chemistry. Proc. Dryland Seep Control, Symp., Edmonton, Canada, p. 2, 1978.

110. Tanji, K.K., L.D. Doneen, G.V. Ferry, and R.S. Ayers, Computer simulation analysis on reclamation of salt-affected soils in San Joaquin Valley, Cal. Soil Sci. Soc. Am. Proc., 36:127, 1972.

111. Jury, W.A., W.M. Jarrel, and D. Dewit, Reclamation of saline sodic- soils by leaching. Soil Sci. Soc. Am. J., 43:1100, 1976.

112. Bingham, F.T. and M.J. Garber, Zonal salinization of the root system with NaCl and Boron in relation to growth and water uptake of corn plants. Soil Sci. Soc. Am. Proc., 34:122, 1970.

113. Wiklander, L., Cation and Anion Exchange Phenomena, Chapter 4 in Chemistry of the Soil, F. Bear (Ed.), Reinhold Pub., New York, pp. 186, 1965. ACS Monograph No. 60.

114. Bower, C.A. and J.O. Goertzen, Replacement of adsorbed sodium in soils by hydrolysis of $CaCO_3$. Soil Sci. Soc. Am. Proc., 22:32, 1958.

115. Balba, A.M., Effect of sodium water and gypsum increments on soil chemical properties and plant growth. Alex. J. Agric. Res., 8:51, 1960.

116. Dielman, Ph., Reclamation of Salt-affected Soils. Land Recln. Inst., Holland Pub., 11:180, 1962.

117. Schoonover, W., M.M. El-Gabaly, and M.N. Hassan, Hilgardia, 1, 1957.

118. Balba, A.M., H. Bassiuni, and H. Hamdy, Use of ^{22}Na in the determination of the gypsum required for ameliorating soil alkalinity. Tal. Agrok., 18:267, 1968.

119. Schoonover, W., Examination of soils for alkali. Agric. Extension Service, USDA, Mimeograph, 1952.

120. Abrol, I.P. and I.S. Dahyga, Flow-associated precipitation reactions in saline-sodic soils and their significance. Geoderma, 11:305, 1974.

121. Doering, E.J. and W.O. Willis, Chemical reclamation for sodic strip-mine spoils. USDA ARS-NC 20, pp. 8, Cited in Rhoades.[89]

122. Prather, R.J., J.O. Goertzen, J.D. Rhoades, and H. Frenkel, Efficient amendment use in sodic soil reclamation. Soil Sci. Am. J., 42:782, 1978.

123. El-Gibaly, H., Gypsum fineness in relation to reclamation of alkali soils. Trans. 7th Cong. Soil Sci., 1:528, 1960.

124. Chawla, K.L. and I.P. Abrol, Effect of gypsum fineness on the reclamation of sodic soils. Agric. Water Manag., 5:41, 1982, Cited in Abrol et al.[28]

125. Keren, R. and P. Xauschansky, Coating of calcium carbonate on gypsum particle surfaces. Soil Sci. Soc. Am. J., 45:1242, 1981.

126. Keren, R. and I. Shainberg, Effect of dissolution rate on the efficiency of industrial and mined gypsum in improving infiltration of a sodic soil. Soil Sci. Soc. Am. J., 45:103, 1981.

127. Khosla, B.K., K.S. Dargan, I.P. Abrol, and D.R. Bhumbla, Effect of depth of mixing gypsum on soil properties and yield of barley, rice and wheat grown in a saline sodic soils. Indian J. of Agric. Sci., 43:1024, 1973. Cited in Abrol et al.[28]

128. Quirk, J.P. and R.K. Schofield, The effect of electrolyte concentration on soil permeability. J. Soil Sci., 6:163, 1955.

129. Hira, G.S., M.S. Bajwa, and N.T. Singh, Prediction of water requirements for gypsum dissolution in sodic soils. Soil Sci., 131:331, 1981. Cited by Rhoades, J.D.[89]

130. Balba, A.M. and A.M. El-Laithy, A laboratory study of the leaching process of saline alkali soil from the north of the Nile Delta. J. Soil Sci., UAR, 8:87, 1968.

131. Verhoeven, B., Leaching of sodic soils as influenced by application of gypsum. Proc. of Salt-affected Soils Symp. Bpst., Hungary, Tal. Agrok., 14:263, 1965.

132. Massoud, F.I., Some physical properties for calcareous soils and their related management practices. FAO Soil Bull. No. 21:73, 1973.

133. Wallace, A. and S.D. Nelson, Forward of the special issue of soil science, 141:311, 1986.

134. Wallace, A. and A.M. Abuzamzam, Interactions of soil conditioners with other factors to achieve high crop yields. Soil Sci., 141:343, 1986.

135. Wallace, A. and G.A. Wallace, Effects of soil conditioners on emergence and growth of tomato, cotton and lettuce seedlings. Soil Sci., 141:313, 1986.

136. Wallace, A., G.A. Wallace, and A.M. Abuzamzam, Effect of soil conditioners on water relations in soils. Soil Sci., 141:346, 1986.

137. Malcolm, C.V., T.C. Swaan, and H.I. Ridings, Nicheseeding for broad scale forage shrub establishment on saline soils. Int. Symp. Salt-Affected Soils. India, 539, 1980.

138. Hamdy, A., Crop management under saline water irrigation, page 108 in Water, Soil and Crop Management Relating to the Use of Saline Water, FAO - AGL/ MISC/16/90, FAO, Rome, 1990.

139. Bernstein, L., L.E. Francis, and R.A. Clark, Interactive effects of salinity and fertility on yields of grains and vegetables. Agron. J., 66:412, 1974.

140. Goldberg, D., B. Gornat, and D. Riman, Drip irrigation principles, Scientific Pub. Israel, p. 296, 1976. Cited in Abrol et al.[31]

141. Amer, F., M.M. El-Gabaly, and A.M. Balba, Cotton response to fertilization on two soils differing in salinity. Agron. J. 56:208, 1961.

142. Balba, A.M. and M.F. Soliman, A quantitative comparative study on the effect of salts on the growth and nutrient uptake by sudangrass. Int. S.S.S. Cong. Edmonton, Proc., 1978.

143. McClung, G. and W.T. Frankenberger, Soil nitrogen transformations as affected by salinity. Soil Sci., 139:405, 1977.

144. Jurinak, J.J. and R.J. Wagenet, Fertilization and salinity, 1981. In Feigin.[146]

145. Sprent, J.E., The effects of water stress on nitrogen fixing root nodules. III- Effects of osmotically applied stress. New Phytologist, 71:451, 1972. In Feigin.[146]

146. Feigin, A., Fertilization of crops irrigated with saline water. In FAO - AGL/MISC/ 16/90, p. 137, 1990.

147. Finck, A., Fertilizers and Fertilization. Verlag Chemic.

148. Lunin, J. and M.H. Gallant, Salinity-fertility interactions in relation to the growth and composition of beans. I- Effect of N, P and R. Agron. J., 57:339, 1965.

149. Sherman, G.D.F., Better crops with plant food. C.A. 46, 5235.

150. Klage, M.G., Effect of clay type and exchangeable cations on aggregation and permeability of solonetz soils. Soil Sci., 102, 46, 1966.

151. Yadav, J.S.P. and P.R. Girdhar, The effect of magnesium to calcium ratios and SAR values of leaching water on the properties of calcareous versus noncalcareous soils. Soil Sci., 131:194, Cited in Salt-Affected Soils and Their Management (Arabic) FAO Soil Serious No. 39, 1989.

152. Kelley, W.P., Alkali Soils, Their Formation, Properties and Reclamation. Reinhold Pub., New York, p. 68, 1951.

153. Martin, J.C., Effect of crop growth on the replaceable bases of some California soils. Soil Sci., 27:123, 1929. Cited in Kelley.[81]

154. Reeve, R.C., C.A. Bower, R.H. Brooks, and G. Schwen, Comparison of the effect of exchangeable sodium and potassium upon the physical condition of soil. Soil Sci. Soc. Am. Proc., 18:130, 1954.

155. Brooks, R.H., C.A. Bower, and R.C. Reeve. The effect of various exchangeable cations upon the physical conditions of soils. Soil Sci. Soc. Am. Proc., 20:325, 1956.

156. El-Ashtar, A.F., A Study of Magnesium Relationships with Soil and Plant. M.Sc. Thesis, University of Alexandria, Egypt, 1976.

157. Rhoades, J.D., Soil salinity assessment. Recent advances and findings. U.S. Sal. Lab., 1993.

CHAPTER 3

Calcareous Soils[*]

INTRODUCTION

A soil is considered "calcareous" from the chemical point of view when it is in equilibrium with excess of $CaCO_3$ at the partial pressure of the atmospheric CO_2. However, still lacking is the definition of this "excess" of soil $CaCO_3$ content at which the properties of the soils as a medium for plant growth and the culture practices for their utilization for agricultural purposes are affected. In this regard the FAO symposium on Calcareous Soils[1] recommended that "the concept of calcareous soils be more precisely defined and the description of these soils be expressed in standardized terminology."

The $CaCO_3$-rich soils are widely spread in the arid regions, where parent rocks prevailing in the region are mostly carbonatious limestone and dolomite or at least rich in calcium as the basalt (Ruellan[2]).

The warm dry climatic conditions of the region which prevail most of the year result in the preservation of $CaCO_3$ in the soil profile. The annual rainfall is not sufficient to dissolve the lime and leach it to the bottom of the soil profile.

MORPHOLOGY

According to Reullan,[2] five characteristics distinguish the soil with a differential calcareous profile.

Form of $CaCO_3$

The $CaCO_3$ of the soil might be present in three different forms:

1. Diffused distribution: $CaCO_3$ exists in fine particles less than 1 mm. It is difficult for the eye to distinguish its existence.
2. Discontinuous, concentrated pseudomycilium in a friable mass or in nodules in several places separated from each other by areas of less $CaCO_3$ content in a diffused distribution.
3. Continuous concentration: The $CaCO_3$ is diffuse or forming nodules and constitutes about 60%.

---

[*] "Calcareous soils" means $CaCO_3$ – rich soils.

Distribution of CaCO₃

These soils are characterized by the presence of three main horizons:

a.   A "B" horizon rich in $CaCO_3$ in the middle of the profile.
b.   Above the "B" horizon another horizon "A" containing less $CaCO_3$.
c.   Horizon "C" also is lower in $CaCO_3$ than horizon "B."

The depth of horizon "B" varies from 0–200 cm. The upper limit is usually clear and its lime content is maximum, but it passes gradually at its base to a horizon of accumulation or nodules of lime.

Texture

The texture of the soils with differentiated profiles varies considerably according to the parent material. Most of these soils have a textural "B" horizon characterized by a layer of clay whose limits are diffuse. Above this horizon, the surface horizon is poorer in clay than the "C" horizon.

Color

Three degrees of darkness in color can be distinguished: dark, light, and very light.

Structure

The structure varies according to the presence of clay and $CaCO_3$ horizons, moisture status, and the depth of profile.

The soil structure more frequent in semiarid and arid regions is angular, less fine, and less stable than in more humid regions. In drier regions, arid and desert regions, the structure is weakly unstable with a fine lamellar structure forming a crust on the soil surface, Reullan.[2]

Mineral Forms

The carbonates in calcareous soils consist of calcium and magnesium carbonates. They are usually evaluated and expressed as carbonate equivalent. The CaO/MgO ratio indicates that calcium carbonate predominates. Also, X-ray diffraction patterns indicate that calcite is the major mineral present in the calcareous soils. Dolomite is usually present in a limited proportion. Aragonite usually constitutes a small fraction (Meester[3]). This may be due to the higher solubility of aragonite than calcite at variable degrees of temperature. Buehrer and Williams[4] postulated that calcium carbonate exists in the soil in forms other than calcite and aragonite as the precipitated form resulting from variable reactions in the soil system.

PHYSICAL PROPERTIES

Particle Size Distribution

It is common practice to remove all the $CaCO_3$ content of soils in the determination of particle size distribution to disperse the soil samples. In highly calcareous soils, this practice is always imperfect. The $CaCO_3$ is not only a cementing agent, but also a distinct component of the mineralogical composition of the soil in various size ranges. Inclusion of $CaCO_3$ in the soil fractions without considering its own size distribution, may result in misleading interpretation of the behavior of calcareous soil when such interpretation is based on particle size distribution alone. Therefore, for highly calcareous soils it is desirable to complement the particle size distribution of the soil with that of its $CaCO_3$. This determination can be achieved by analysis of the whole soil sample with and without removal of its lime content. Menon's[5] results in the El-Ghab Project, Syria (Table 18), as stated by Massoud,[6] showed that the texture of the soil would be clay loam if determined in the laboratory without removing $CaCO^3$, although the clay and silt fractions are those usually found in a noncalcareous loamy sand or sandy soils. The same soil would Probably be classified as silt clay by a soil surveyor in the field. The data in Table 18 also show that the noncalcareous clay and silt fractions amount to 20 percent compared to a corresponding lime content of 51% which is very high and creates special fertility problems. To avoid destruction of the soil particles, the decarbonation must be carried out with a very dilute acid. For a given size fraction, the difference between these two analyses gives the amount of $CaCO_3$ present in the size range.

Soil-Moisture Relationship

The moisture tension drying curve is of practical significance in the determination of the range of soil moisture available for plant growth and its depletion pattern. Massoud et al.[7] stated that the moisture characteristics of the highly calcareous soils of the northwest coast of Egypt showed the shape of the moisture curves of the sandy soils. There was a marked decrease in the moisture content with increasing tension up to 1.0 atm as compared to that at a higher tension. El Gabaly[8] stated that wilting occurs at 9–10%, the water-holding is at about 19–21%, and the available water range is not more than 10–12%. He added that from various studies it has been found that most of the available water is utilized before or even at a moisture tension of 1 atm, while in the Delta clay loam soils, this may occur at a tension of 4 atm. Since most of the lime is in the silt fraction, one would expect that the ability of these soils to retain moisture would be rather limited. The studies have revealed that most of the water is being held by physical forces and that is why the decrease in the available moisture occurs rather abruptly and not gradually as is the case in the alluvial soils of the Delta containing about 28 $CaCO_3$. These moisture characteristics are related to the efficiency and frequency of irrigation.

Water Movement Through the Soil

It has been noticed that mobility of water through the calcareous soils is fast. This fast movement of water increases salinization hazards from groundwater tables or water streams that pass adjacent to the soil. Balba and Soliman[9] showed that the groundwater reached the surface of columns of calcareous loam, sandy and clay soils 50 cm long after

Table 18.   Soil particle and $CaCO_3$ equivalent size distribution of Pilot Area, El-Ghab Project, Syria.

	Clay <0.002 mm %	Silt 0.05–0.002 mm %	Sand 0.05–2.0 mm %	$CaCO_3$ %
Before removing carbonates	28	43	29	
After removing carbonates	9	11	12	68
Distribution of carbonates	19	32	17	
Percent loss due to removing carbonates	68	74	60	
Distribution of carbonates as percent of total $CaCO_3$ equivalent	28	47	25	

Source: Massoud.[6]

1, 16 and 28 days, respectively. Accordingly, the calcareous loamy soil columns accumulated 632.35 meq Cl⁻ after 125 days, while the clay and sandy soil columns accumulated 442.77 and 319.12 meq, respectively.

The diffusivity is a soil property which can be defined when the moisture content "Q" is specified (Childs and Collis-George[10]) and can be calculated from the relationship:

$$D = k (\partial Y / \partial \theta) \tag{29}$$

where $\partial Y / \partial \theta$ =   the inverse of the specific water capacity at a particular value of θ.
        k =   the hydraulic conductivity.

The diffusivity of water through calcareous soils was found to be higher than through noncalcareous soils of similar texture. Increasing $CaCO_3$ up to 10–15% increased diffusivity, while a further increase up to 20 or 25% relatively reduced it. In highly calcareous soil, it seems that size distribution of the carbonate had more effect on diffusivity than did its total content, Massoud.[6]

Diffusivity, hydraulic conductivity, and initial moisture content affect water infiltration and distribution of moisture in the soil. These properties are of significant importance to irrigation practices and conservation of soil and water. In calcareous soils containing a crust on the surface or a hardpan within the profile, the water infiltration rate would slow down.

Crust Formation

The formation of crusts is a problem in the carbonate-rich soils newly put under cultivation. Crusting which takes place at the soil surface hinders seedling rate of emergence and percentage. The adverse effect of crusts depends on their strength and thickness.

Hillel[11] stated that Richards[12] found that emergence of bean seedlings in fine sandy loam soil was reduced from 100 to 0% when crust strength increased from 108 to 273 mbar, whereas Allison[13] reported that emergence of sweet corn was prevented only when crust strength exceeded 1200 mbar.

The calcareous soils slake into a dispersed mass upon saturation with water. The dispersed surface layer clogs the surface macropores, thus infiltration of water into the soil and aeration tend to decrease (Hillel[11]).

Lemos and Lutz[14] showed that increase in the silt or the soil fractions less than 0.1 mm, the 2 : 1 type clay mineral, compaction by rainfall and soil puddling increase crust strength. Fuller and Padgett[15] showed that the calcareous soils they used became severely crusted and compacted after repeated flooding. They stated that the action of water disperses the soil and reorients the soil particles. Thus, a close compact arrangement resulted when the soil was dried. According to Lemos and Lutz,[14] Fuller and Padgett,[15] and Massoud,[6] wetting and drying and application of organic materials decrease crust formation. Massoud[6] recommended that the frequency of irrigation should be close enough to prevent drying of the soil surface and hardening of the crust.

The mechanism of crust formation in calcareous soils seems to follow a sequence of processes involving slaking and break down of aggregates, segregation, dissolution of $Ca(HCO_3)_2$, rearrangement of particles, and redeposition and cementing by $CaCO_3$ on dessication. Massoud[6] observed that the thickness of the crust increases five-fold by increasing the water application from 0.3 to 5 field capacity.

Reellan[2] described the types of crustation in calcareous soils as follows:

• Non platy lime crusting.
• Platy, the crust is usually surrounding a non platy lime crusting.
• A compact slab on a crust and on a non platy lime crust.

CHEMICAL AND FERTILITY PROPERTIES

CaCO₃-Water Suspensions

Though calcium carbonate is a simple chemical compound it is found in nature in variable forms different in their physical and chemical properties. The soil calcium carbonate which was formed as a result of very slow geological processes have different properties from the friable fine particles of carbonate which are formed by precipitation in the soil system, as a result of reaction between $Ca(HCO_3)_2$ or HCO_3 and the free alkalis in the soil. The main differences are their degree of solubility in water and water hydrolysis.

The main forms of calcium carbonate in nature are aragonite and calcite.

Aragonite is more soluble in water than calcite at variable degrees of temperature. Accordingly, calcite is more stable and is more frequently encountered in soils. However, aragonite might be converted into calcite. The aragonite contains 2 OH groups in every 63 molecules of $CaCO_3$. It contains only simple $CaCO_3$ molecules, while calcite contains one double molecule of $(CaCO_3)_2$ for every 8 simple $CaCO_3$ molecules.

Aragonite is more hydrolyzable in water. The pH of 1:25 and 1:50 aragonite-water suspensions was 9.82 and 9.77, respectively. The pH values of the same calcite-water suspensions were 8.96 and 9.28, respectively (Buehrer and Williams[4]).

The precipitated $CaCO^3$ formed from $CaCl_2$ and $(NH_4)_2CO_3$ is the only form which has a chemical composition close to the pure $CaCO_3$. The resultant precipitate is granular and the soluble salts which may be present are easy to remove. On the other hand, the precipitate resulting from $Ca(OH)_2$ and CO_2 was free of adsorbed impurities.

In some soils, calcium carbonate may be coated with soil organic or inorganic colloids, which decreases its hydrolysis with water or its reaction with CO_2. Buehrer and Williams[4] equilibrated calcium carbonate with various concentrations of dialyzed ferric hydroxide, starch, agar, gelatine, or clay. They found that the pH of each system was generally lower than that of $CaCO_3$ alone (Table 19). Most studies agree that complex compounds and soluble ion pairs constitute a significant proportion of $CaCO_3$ solution (Nakayama[16]).

In calcareous soils systems, although $CaCO_3$ is a major factor controlling the properties of the systems, it is also affected by other soil characteristics such as the pH of $CaCO_3$-water suspensions. Also, colloidal clay causes the pH of $CaCO_3$-water suspension to decrease.

Precipitated calcium carbonate in the presence of NaCl solution at controlled CO_2 partial pressures resulted in calcite as detected by the X-ray diffraction, while in the presence of $MgCl_2$, the calcite and aragonite were detected and Mg was coprecipitated in the solid phase. In solution of Na_2SO_4, the solubility of the precipitate was higher than that in the case of $MgCl_2$ solution and both calcite and natrite were detected. In $MgSO_4$ solution, the solubility of the precipitate was intermediate between that in the presence of $MgCl_2$ and of Na_2SO_4 solutions (Doner and Pratt[17]).

$CaCO_3$-Soil Systems

The presence of $CaCO_3$ in soils imposes an unescapable problem in the study of any reaction in these soils. When in equilibrium with CO_2 of the atmosphere, calcareous soils with free $CaCO_3$ will have a pH of 8.2–8.4. With $MgCO_3$ $3H_2O$, the pH values are about 1.5 unit higher than the same conditions with $CaCO_3$. Seatz and Peterson[18] pointed out that because of the reaction of $CaCO_3$ hydrolysis,

$$CaCO_3 + H_2O \rightarrow Ca^{++} + HCO_3^- + OH^-$$

and other relationships between moisture and pH, the control of excess water seems to be important in calcareous soils. The build-up of hydroxyl ions from the hydrolysis of $CaCO_3$ is prevented by adequate CO_2 pressure. Thus moist calcareous soils may become excessively alkaline. In $CaCO_3$-rich soils, the dominant weatherable mineral is calcium carbonate. In these calcareous systems, the important rate limiting step in the dissolution kinetics is the transfer of atmospheric CO_2 to solution (Amrhein et al.[19]). They modeled the dissolution of calcite by three simultaneous reactions on its surface:

a) an adsorption of CO_2 (eq.) on the calcite surface followed by
b) hydrolysis of the CO_2 to H_2CO_3, $CO_2 + H_2O - H_2CO_3$
c) the hydrolysis is catalyzed by the calcite surface, and occurs slowly.

Clark[20] found that measured solubility of calcium carbonate in soils was affected by:

•   errors in the liquid functions in the measurement of soil pH.
•   slow attainment of equilibrium.
•   complexing of calcium by sulfate and other anions.

Clark[20] concluded that after minimizing the pH-meter junction effects and correcting for sulfate complexing, the PKsp value for the calcium carbonate was found to be within 0.2 unit of that of calcite.

Table 19.   Effect of colloidal soils on the pH of suspension of precipitated calcium carbonate.

Colloid	% Colloid						Initial pH of Soils	$CaCO_3$ pH
	1.0	0.5	0.10	0.05	0.01	0.001		
Agar	—	—	6.84	—	7.64	7.88	5.85	9.98
Starch	—	7.77	7.90	—	8.70	8.96	7.80	9.98
Gelatin	—	8.78	8.80	—	8.89	8.78	8.63	9.98
Fe(OH)$_3$	7.46	—	8.78	—	9.7	9.79	6.22	9.98
Clay	8.05	8.05	8.54	8.28	—	—	8.16	9.98

Ratio of $CaCO_3$ to soil = 1:10.
Source: Buehrer and Williams.[4]

Balba[21] had shown that Ca^{++} from the dissolved soil $CaCO^3$ had participated in the exchange reaction in the soil system. The participation of $CaCO_3$-Ca raised the point that calcareous soils may resist deterioration when irrigated with saline water. Balba[21] showed that use of saline noncarbonated water for irrigating calcareous soils dissolved portions of the soil $CaCO_3$. The dissolution ceased when the saline water contained residual carbonate and consequently, the soil exchangeable sodium percentage (ESP) increased.

Also, improvement in soil physical properties, when soils high in free alkali carbonates are treated with Ca amendments, may not be achieved as rapidly as expected because a part of the dissolved Ca may be in a complexed form, and consequently does not participate in the exchange reaction (Nakayama[16]).

Solubility of Soil $CaCO_3$

$CaCO_3$ is sparingly soluble in CO_2-free distilled water at room temperature. Its solubility is controlled by the soil pH, the salts present in the soil system, the amount of the organic matter, the amount of applied water, and the frequency of its application.

In the alkaline soils, the $CaCO_3$ solubility decreases with the increase in the pH of the soil system (Table 20) (U.S. Sal. Lab.[22]).

Neutral soluble salts favor the solubility of the soil $CaCO_3$ provided they do not have a common ion with $CaCO_3$. Thus, NaCl favors the $CaCO_3$ solubility, while $CaCl_2$ suppresses it.

The decomposition of the soil organic matter produces CO_2 which dissolves $CaCO_3$. The effect of water on $CaCO_3$ hydrolysis was mentioned above.

Temperature also has a significant effect on dissolving $CaCO_3$ under hot climates.

Cation Exchange Capacity (CEC)

The CEC of the calcareous soils depends on their clay and organic matter type and content. In areas of Egypt known as highly calcareous, west of the Nile Delta, the CEC is about 15 meq/100 g soil. Because the nutrient elements in the exchangeable form are considered available to plants, the greater the CEC of the base-saturated soils, the higher their fertility.

Table 20.   CaCO$_3$ solubility in different pH values.

pH	Sol. CaCO$_3$ meq/l	pH	Sol. CaCO$_3$ meq/l
6.21	19.3	8.60	1.1
6.50	14.4	9.20	0.82
7.12	7.1	10.12	0.36
7.85	2.7		

Source: U.S. Sal. Lab. Handbook No. 60.[22]

The Nutrient Elements

Nitrogen

Amounts and Transformation

The organic matter (OM) is the main source of the mineral forms of soil nitrogen. In regions with arid warm climates low in "OM," the soil fertility in nitrogen is low.

The nitrogenous fertilizers are subject to chemical reactions and biological activities when they are incorporated in the soil. Such reactions and activities might affect the form of the added fertilizer and consequently its fertilization value. Balba et al.[23] had shown that the capacity of the soil of the northwest coast of Egypt rich in CaCO$_3$ to adsorb and "fix" the added NH$_4$ is lower than that of the alluvial clay soil. Accordingly, a greater part of the added NH$_4$ is left in the soluble form. The greater the amount of soluble NH$_4$ the more is it subject to loss by volatilization. A test of soil microorganisms' activities was carried out by estimating the percentage of nitrified NH$_4$ upon its incubation for 3 weeks in the soil by Balba et al.[23] They showed that 59.4% of the added (NH$_4$)$_2$SO$_4$ was transferred to nitrate in the calcareous northwest coast soil. The corresponding percentage in the alluvial soils was 71.2%.

Loss of NH$_3$ by Volatilization

It has been known since Sprengel in 1839 and Boussingault in 1849 that when ammonium fertilizers are added to calcareous soils, a part of ammonia is lost by volatilization. More recently, several investigators (Terman and Hunt[24]; Balba and Nasseem[25]; Fenn and Kissel[26,27]; Stumpe et al.[28]) have shown that the volatilization of NH$_3$ from fertilizers applied to soils is affected by several factors.

Balba and Nasseem[25] showed that the lost amount of NH$_3$ differed from one soil to another, with the calcareous soil being the most effective in NH$_3$ volatilization.

Table 21 shows that addition of CaCO$_3$ to the soil which contained initially 5.6% CaCO$_3$ increased the loss of NH$_3$. When the soil content of CaCO$_3$ was removed by dilute HCl treatment, the loss of NH$_3$ decreased to 2.2% of the applied amount of NH$_4$$^+$. Table 21 also shows that increase in the soil pH due to increments of CaCO$_3$ was slight though the loss of NH$_3$ increased. This result invalidates the effect of pH as a sole factor in the loss of NH$_3$ in calcareous soils. The effect of CaCO$_3$ is mainly due to the formation of (NH$_4$)$_2$CO$_3$ which is easily decomposed releasing NH$_3$ as follows:

Table 21.   The effect of $CaCO_3$ on the loss of NH_3 from ammonium sulfate.

% $CaCO_3$	N Added mg/100 g Soil	pH[1]	N Loss of Added
0.0	100	6.0	2.2
4.6	100	8.4	24.5
6.0	100	8.5	27.5
9.0	100	8.6	31.5
14.0	100	8.6	34.0
25.0	100	8.6	42.0

[1]   pH in 1:2 soil water suspension.
Source: Balba, Nasseem, and Massry.[23]

$$(NH_4)_2SO_4 + CaCO_3 \rightarrow CaSO_4 + (NH_4)_2CO_3$$

The decomposition of $(NH_4)_2CO_3$ giving off $CO_2 + NH_3$ drives the reaction to the right side, otherwise this reaction would not proceed.

Fenn and Kissel[26,27] showed that NH_3 loss in gaseous form from $(NH_4)_2SO_4$ applied to soils containing 1.3% $CaCO_3$ greatly increased as compared to the loss in the presence of 0.5% $CaCO_3$. They stated that the N losses were about 25% of the amount applied regardless of temperature. They established two multiple regression equations to evaluate NH_4-N losses with respect to time, temperature, rate of NH_4-N application, and soil $CaCO_3$ content. The correlation coefficients (R^2) were +0.88 and +0.85.

Form of N-Fertilizers

Balba and Nasseem[25] had shown that the loss by volatilization was from $NH_4OH >$ $(NH_4)_2SO_4$ (AS) > urea > NH_4NO_3 (AN). The loss from $Ca(NO_3)_2$ was nil since it does not contain or form NH_4. Loss of NH_3 from (AS) was greater than from (AN) on the basis of NH_4 content. This is explained by the reaction between (AN) and (AS) with $CaCO_3$.

$$\text{With AN: } 2NH_4NO_3 + CaCO_3 \leftrightarrow Ca(NO_3)_2 + (NH_4)_2CO_3$$

The $Ca(NO_3)_2$ formed is soluble and an equilibrium is more likely to take place.

$$\text{With AS: } (NH_4)_2SO_4 + CaCO_3 \leftrightarrow CaSO_4 + (NH_4)_2 CO_3$$

The solubility of $CaSO_4$ is low and the reaction is more likely to be driven to the right side (Terman and Hunt[24]).

In the case of NH_4OH, the loss of NH_3 was extremely high. The pH values of the soil-NH_4OH-system were 9.2 and 10.2 for the systems used in the study.

Stumpe et al.[28] showed that loss of NH_3 by volatilization from surface applied urea ranged from 10 to 308 of the applied N dependent on soil type. Addition of urea with phosphate (100 g N kg^{-1}) slightly decreased NH_3 loss from 17.0% to 12.2% of the applied N on a highly calcareous (225 g $CaCO_3$ kg^{-1}). Mixtures of urea and ureaphosphate were substantially less effective in reducing NH_3 volatilization with relative reductions ranging from 0% to 20%. They concluded that urea-ureaphosphate mixtures should probably not be recommended as a means of reducing NH_3 volatilization on highly calcareous soils, though its use on noncalcareous soils and soils low in $CaCO_3$ merits further testing.

Loss of N from Calcareous Soils by Leaching

Loss of N by leaching from the $CaCO_3$-rich soil is controlled by its texture. The form of N usually leached is the NO_3^- form as it is not adsorbed or immobilized in the soil.

Phosphorus

The soil phosphorus assumes several chemical forms depending on the parent material and the factors of soil formation prevailing in the region. In calcareous soils, $CaCO_3$ has a dominating influence due to its properties of relatively moderate solubility, high buffer capacity, and alkalinity. Under such conditions the predominant phosphates are the inorganic phosphates which constitute as much as 90–95% of soil total phosphates. Organic P forms constitute the remaining 5–10% of total soil P. Using a sequence of extractants which selectively remove phosphorus forms one after another, it was shown that calcium phosphates are the prevailing soil form while the Al-P and occluded forms are of the order 10–11% of total calcareous soil P content (Solis and Torrents[29]).

Phosphate Reactions with CaCO₃

Boischot and Herbert[30] at Versailles Experimental Station, Paris, France, showed that the amount of P which disappeared from solution had increased with increasing P concentration to the system containing a constant amount of $CaCO_3$. They concluded that in low phosphorus concentration, $CaCO_3$ sorbs P on its surface, but in high P concentration, precipitation occurs. Cole et al.[31] repeated Boischot and Herbert's experiment by increasing P concentration with the amount of $CaCO^3$ constant. They also showed that reversion of P had increased with the increase of soluble P concentration in the system. To test for the sorption mechanism of $CaCO^3$, they applied the Langmuir adsorption isotherm by plotting:

$$\frac{\text{P concentration in equilibrium solution}}{\text{mg P sorbed / g } CaCO_3}$$

against P concentration in the equilibrium solution. Their results showed that the above relationship is rectilinear up to a P concentration of 3×10^{-4} M. Beyond this concentration, the Langmuir isotherm did not hold, indicating that another reaction must have taken place. Also, the phosphate sorption at a given level in the lower range of concentration was directly proportional to the amount of $CaCO_3$ present, indicating simple surface sorption of phosphate. Using radioactive ^{32}P, Cole et al.[31] obtained direct evidence confirming the surface sorption process. All of the phosphate sorbed in the lower range was exchanged with the radioactive phosphorus in the solution. On the other hand, less than one third of the phosphate sorbed at higher concentrations was isotope exchangeable. At higher phosphate concentrations it was found that the amount of P sorbed was not a function of the $CaCO^3$ surface and the reaction proceeds until the phosphate concentration of the solution had dropped below a certain critical value indicating the precipitation of some form of calcium phosphate.

Soils rich in exchangeable Ca may retain HPO_4^- on the clay surfaces through the exchangeable Ca:

$$Ca \qquad\qquad\qquad Ca - HPO_4$$

$$Ca + 2H_2PO_4^- \;\rightarrow\; Ca - HPO_4$$

Thus in calcareous soils rich in $CaCO_3$ and exchangeable Ca, phosphorus may be immobilized by any or all the following mechanisms:

a. Adsorption on active sites of $CaCO_3$.
b. Precipitation by Ca in the system.
c. Reaction with the exchangeable Ca.

The $CaCO_3$-rich soils usually have alkaline pH. Calcium phosphates under pH values above 8.0 are usually precipitated forms such as tricalcium phosphate or hydroxy apatite (Balba[32]).

The $NaHCO_3$ method introduced by Olsen et al.[33] for testing soil fertility in phosphorus is widely used in many laboratories in regions of $CaCO_3$-rich soils. However, there is a need to estimate the soil P supply to the growing plants in addition to the soil test value determined before planting. This estimate can be obtained by calculating the capacity factor which indicates the amount of P from the solid phase that will enter the soil solution either by. (1) determining the labile P measured by ^{32}P exchange, (2) using anion resin to determine extractable P, (3) calculating Beckette and White[34] potential buffer capacity "PBC" from the amount of soil P "Q" and the concentration of soluble soil P "I" or, Q/I or (4) calculating the differential potential buffering capacity "DPBC" of Jensen.[35] Jensen, cited by Olsen,[36] defined "DPBC" as the amount of P to be added or removed per 1 g of soil in order to obtain a certain alteration of the phosphate potential (0.5 p Ca+p HPO_4).

Potassium

Soils under dry climates are generally rich in potassium because leaching is not effective. This is especially true when the soil parent material contains potassium. Several primary silicate minerals are rich in K such as muscovite, biotite, and potash feldspars. Mica minerals are main sources of soil K. Also, the hydrated mica-like clay minerals such as illite are an indication that the soil is rich in total K.

The Exchangeable Potassium

This form is the most important soil form of potassium to plant nutrition though it is usually present in relatively small amounts. Calcareous soils contain about 0.5 to 1.5 meq

K per 100 g while soils of humid regions contain about 0.15 to 0.5 meq K/100 g. In many laboratories water soluble plus exchangeable soil K—the labile K—are determined together as an indication of soil fertility in K.

Nonexchangeable Potassium

By nonexchangeable K it is meant all soil potassium forms that are not extractable with the salt solution used for extracting the exchangeable form including the "fixed" potassium and the mineral K. Nonexchangeable soil K constitutes the major portion of soil K. An equilibrium between soil exchangeable plus soluble K and nonexchangeable K usually exists in the soil.

The chemical determination of K in soil samples from Egypt is presented in Table 22. Balba[37] indicated that the calcareous soils contained significant amounts of potassium. The samples No. 4, 5, 6, 7, and 9 from the Nile Delta were not richer than the calcareous soils No. 1, 2, 3, 8, 10, and 11. The HCl-extractable soil K was of the order of 10 meq/100 g soil and the soil exchangeable K plus soluble K forms was about 1.0 meq per 100 g soil.

Positive response to K-fertilizer application rarely takes place and depends mainly on the kind of crop.

Micronutrients

The $CaCO_3$-rich soils are known for their low fertility level of micronutrients. Several factors are responsible for this low level, mainly the alkaline pH, low proportion of clay, and in most cases low organic matter content.

Amounts and Forms of Micronutrients

Iron forms of 36 soil samples representing regions in Egypt were as follows (Abdel Kader and Abu Ghalwa[38]):

- The total iron content was the highest (12.9%) in the clay soils, while the calcareous soils contained the lowest total Fe percentage (1.8–5.6%).
- The clay and the $CaCO_3$-rich soils contained 0.8–1.8% free iron.

Total Zn content of the $CaCO_3$-rich soils was in the range 33–55 μg g^{-1}, while that of the clay alluvial soils ranged between 80–112 μg g^{-1} (Kishk and Mohamed[39]). They stated that percent organic matter and cation exchange capacity accounted for 90% of variations in total soil Zn.

In order to obtain an estimation of the soil fertility in microelements, several extracting solutions were used assuming that these solutions extract definite forms of the elements studied. The main solutions are NH_4-acetate of pH 4.0 or 7.0, Na EDTA, EDTPA, and hydroquinone to extract the soluble + exchangeable, the chelated, and the easily reducible forms of the elements, respectively. Other solutions are also used but not as common as the above mentioned ones.

The effect of adding organic matter on the transformation of soil native Mn and on the MnO_2 applied to the soil depended on the kind and level of organic material, time of incubation, and soil characteristics. Generally, a temporary increase in the water-soluble

Table 22. Exchangeable, HCl-extractable, and Neubauer values of potassium in soil samples from different locations.

Soil	Exch. K meq/100 g Soil	Neubauer Values meq/100 g[1]	HCl-Extractable meq/100 g Soil
N.W. Coast (calc.)	0.94	0.85	12.1
N.W. Coast (calc.)	1.04	0.60	11.4
N.W. Coast (calc.)	0.82	0.51	8.0
Khazan	0.82	0.8	6.8
Kafr El Dawar	1.80	1.71	13.3
Insha	3.7	1.80	15.0
Kafr Khadr	1.87	1.7	16.0
Maryut (calc.)	1.0	—	—
Gimmezah	1.0	—	—
Rafah (calc.)	1.0	—	8.0
Hammam (calc.)	0.34	—	11.34

[1] K in the plants grown in washed sand was subtracted.
Source: Balba.[37]

plus exchangeable Mn was observed and this effect was more pronounced in the soil initially low in this Mn fraction. The increase in this form was accompanied by a decrease in the easily-reducible Mn even though the relationship was not shown to be quantitative. The behavior of Na_2 EDTA-extractable Mn was different. Its value was closely related to the formation of organic material decomposition products which are capable of maintaining Mn in the insoluble chelated form (Ghanem et al.[40]).

The same trend was manifested when MnO_2 was added to the soils and incubated with the organic materials.

Balba and El Khatib,[41] Thabet,[42] and Alwash[43] had shown that decreasing the soil pH increased its soluble content of Fe, Mn, and Zn. On the other hand, incubation of the soil sample with Na_2CO_3 decreased the extractable amounts of Fe, Mn, and Zn. Increments of $CaCO_3$ to the soil resulted in a decrease in the extractable amounts of Fe, Mn, and Zn. Additions of neutral soluble salts did not affect the status of the three nutrients.

Iron Transformation and Availability to Plants

Halverson and Starkey[44] studied the transformation of iron from Fe_{II} to Fe_{III} in pure solutions. After discussing the factors involved, they stated that conditions of the soil system differ from solution cultures because the iron remains in organic soluble forms under aerobic conditions and under pH values close to neutrality. In solution cultures both Fe_{II} or Fe_{III} almost do not exist at pH 7. Fe_{II} concentration at pH 7 is 27×10^{-6} and Fe_{III} is 61×10^{-11} mg kg^{-1}. They summarized the transformation of iron in the soil by stating that iron solubility in nature depends on the equilibrium conditions which are controlled with partial pressures of oxygen and carbon dioxide and the concentration of active hydrogen and the organic compounds. These equilibrium conditions may vary considerably with the soil microbial activity.

Iron availability to plants under reduction conditions increases because Fe_{II} is more soluble under these conditions. On the other hand, the plants are less capable to uptake nutrients under anaerobic conditions because of lower oxygen in the root zone.

McGeorge[45] studied the causes of chlorosis of plants grown on calcareous soils. He concluded that:

1. Chlorotic seedlings absorbed more calcium and less potassium than nonchlorotic plants though Ca:K ratio in the chlorotic plants' leaves was not much disturbed. This ratio, however, was disturbed in the roots.
2. The increase in absorbed Ca was accompanied with an increase in Fe absorption though most of the absorbed iron was inactive iron (insoluble in HCl) and mostly tied in the roots.
3. Addition of $CaCO_3$ to the acid soils decreased the active iron in the seedlings and increased Ca.
4. Seedlings grown in continuous darkness contained less active iron than those grown under day light.
5. Application of sulfur or a mixture of farmyard manure and sulfur increased active iron but did not increase its absorbed amount.

The problem of lime-induced chlorosis, which takes place on some plants grown on calcareous soils, has been a controversial issue. Several studies have been carried out which led to several postulations, including the following:

a. The calcareous soils contain amounts of $CaCO_3$ which distinctly affect the soil properties related to plant growth, whether they are physical such as soil-water relations and crusting or chemical such as availability of plant nutrients, El-Gabaly.[8]
b. Inskeep and Bloom[46] concluded that their study showed a correlation between increases in soil solution HCO_3^-, total $CaCO_3$, and soybean chlorosis. They considered the bicarbonate (HCO_3^-) as the most important factor for Fe chlorosis of dicotyledonous plants growing in calcareous soils.
c. The degree of chlorosis is related to the calcite ($CaCO_3$), Bureau[47]; Loeppert et al.[48]
d. The presence of $CaCO_3$ does not guarantee the incidence of chlorosis nor does the quantity of $CaCO_3$ always correlate well with chlorosis within a given calcareous field.
e. The reader's attention should be brought to the fact that the fertility problems in calcareous soils are mostly independent of total free $CaCO_3$ in these soils. As examples, the soils used by McGeorge[45] in his study of lime-induced chlorosis contained about 1 to 4% $CaCO_3$. The trip, Quinlan, and Mellville soils used by Brown et al.[49] and by Olsen[36] contained 0.2–0.5, 2.0, and 9.0% $CaCO_3$, respectively. The Hacienda loam and the Chino clay soils contained 32.0 and 40.0% $CaCO_3$, respectively. The alluvial soils of Egypt contain about 3.0% $CaCO_3$ but chlorosis does not appear on plants grown on these soils.
f. Kadry[50] gave much emphasis to the active fraction of $CaCO_3$ which has a high specific surface area.
g. Hartwig and Loeppert[51] consider that the reactivity of carbonate phases strongly influences certain reactions of calcareous soils such as carbonate dissolution kinetics, rhizosphere pH, and adsorption reactions.

The carbonate reactivity in calcareous soils depends on several physical, chemical, and mineralogical factors among which are:

• the mineral composition.
• surface composition and morphology of the carbonate mineral phase.

- particle size distribution.
- degree of aggregation of the carbonate-phyllosilicate complex.
- the accessibility of carbonate mineral surface sites to the soil solution.

Fuehring[52] concluded that when water and nutrients can be supplied, the inherent productivity of calcareous soils over large areas is excellent due to favorable growing season conditions. The micronutrients tend to be less available with increasing pH, but the occurrence of deficiency is highly erratic.

Effect of HCO_3^- and CO_3

Several investigators have shown that Fe accumulation in the shoots is reduced when plants are grown in nutrient solutions containing bicarbonate. The respiration of root tips was reduced in its presence for plants susceptible to lime-induced chlorosis (Miller and Thorne[53]). The cytochrome oxidase activity decreases in HCO_3^- presence (Miller et al.[54]) and the capacity of roots to reduce Fe_{III} to Fe_{II} decreased by increasing pH or introducing a HCO_3^- concentration to the nutrient solution.

The role of HCO_3^- in lime-induced chlorosis of susceptible plants has been a controversial issue. Brown et al.[49] stated that bicarbonate has an indirect action in Fe chlorosis. Coulomb et al.,[55] as stated by Brown and Jolley,[56] found that young leaves of soybeans did not increase in foliar P with added HCO_3^- and concluded that this result does not support the indirect HCO_3^- action. Also, Mengel et al.[57] found no relationship between Fe chlorosis on grapes and the HCO_3^- content of the soil.

Loeppert[58] presented the reaction between Fe_{II} and Fe_{III} salts with $CaCO_3$ in oxidizing environments to form solid phase Fe oxides as follows:

$$4Fe^{2+} + O_2 + 4CaCO_3 + 2H_2O \rightarrow 4FeOOH + 4Ca^{2+} + 4CO_2$$

$$2Fe^{3+} + 3CaCO_3 + 3H_2 \rightarrow 2Fe(OH)_3 + 3Ca^{2+} + 3CO_2$$

The bicarbonate concentration in the calcareous soil solution is controlled by several factors:

a. The equilibrium between soil carbonates and soil atmospheric carbon dioxide.
b. Poor drainage which may cause high partial CO_2 pressure.
c. High soil-water content.
d. Soil composition.
e. High microbial respiration rates.

From the investigations of Lindsay and Thorne,[59] Miller and Thorne,[53] Miller and Evans,[60] Brown et al.,[49] and Romheld et al.,[61] Loeppert[58] concluded that "Bicarbonate-induced stress in nutrient culture is a real phenomenon." However, the discussion persists concerning the bicarbonate effect which may be attributed predominantly to:

1. the influence of bicarbonate on concentration of dissolved Fe in the nutrient solution,
2. a direct effect of bicarbonate on Fe transport or metabolism by the plant, or
3. the influence of bicarbonate on Fe-stress response mechanism.

The anion HCO_3^- is transformed in the soil system to CO_3^-. Also, CO_3^- can be transformed in the soil to HCO_3^-. The influence of this anion on reactions which take place in the soil system is far-reaching. It raises the soil pH because its alkali and alkaline earth salts dissociate giving alkaline reaction. The alkaline pH as well as the CO_3^- anion precipitate several cations among which are the micronutrients. In the $CaCO_3$-rich soils, it is the CO_3^- and not the Ca^{++} that causes the nutritional and physiological disturbances in the growing plants (McGeorge[45]).

The extracted forms of Fe, Mn, and Zn were decreased when the soil was incubated with Na_2CO_3 solution as stated in Chapter 2 and presented in Table 12. Using four alluvial soils containing 3–6% $CaCO_3$, El-Khomsy[62] added increments of $CaCO_3$ and extracted Fe, Mn, and Zn using different extractants. Table 23 shows the average of the results of the four soils. The extracted amounts of Fe, Mn, and Zn decreased with the increase in the added $CaCO_3$.

The decrease in the extracted Fe, Mn, and Zn forms which are considered available to plants was reflected on the absorbed amounts by plants of these nutrients (Table 24). Balba and El Khatib[41] and Thabet[42] showed that corn plants irrigated with water containing 3 meq Na_2CO_3/l absorbed lower amounts of Fe and Mn than the plants irrigated with tap water. The root:shoot ratio of Fe in the untreated plants was 0.62 and increased to 2.10 in the Na_2CO_3-treated plants. The corresponding ratios for Mn were 0.23 and 0.39, respectively.

The nutritional disturbance was expressed by a lower rate of growth of corn plants. Without CO_3^-, the corn plants 30 days old grew at a rate of 35.2 mg per day per plant. When CO_3 was incorporated, this rate dropped to 16.7 mg/day/plant. The uptake of Fe per day per plant of corn, accordingly, decreased from 7.7 μg to 3.7 μg in the CO_3-enriched soil. In a similar study of the rate of Mn and Zn uptake by corn plants 15 days old, this rate decreased from 1.09 and 0.74 μg/day/plant to 0.83 and 0.47 μg/day/plant in the CO_3-enriched soil for both nutrients, respectively.

Balba et al.[63] determined the ratio and amount of Fe^{++} and Fe^{+++} iron in different parts of 20- and 40-day-old beans *(Vicia faba)* grown in pots under normal and HCO_3-treatment. The Fe^{++} and Fe^{+++} iron concentrations in the plant varied according to its age, the plant part, the order of leaves, and HCO_3-treatment. They found that dry weight of the whole plants or of plant parts were not affected by the irrigation with HCO_3-water in the first 20 days. Plants 40 days old were decreased in weight and chlorosis appeared on the leaf underneath the bud in the case of irrigation with HCO_3-water.

It may be concluded that plants growing on soils rich in $CaCO_3$ are likely to suffer from Fe, Mn, and Zn deficiency due to:

a.   decrease of the soil forms of these elements which are available to plants, and
b.   absorbed amounts of Fe, Mn, and Zn are less mobile and their translocation to the upper plant parts is slowed.

Worth mentioning is that gypsum did not affect neither the amount of Fe and Mn extracted nor the absorbed by wheat and beans (Kandil[64]).

Using radioactive ^{59}Fe, Balba and Osman[65] showed that the percent of Fe derived from applied Fe by wheat plants increased with the increase of the Fe rate of application. Increments of $CaCO_3$ decreased the efficiency of applied Fe. This is an indication that Fe applied to calcareous soils is less efficient than Fe supplied to the growing plants. Only high rates of Fe application could restore the Fe concentration in the plants to its original concentration (Table 25).

Table 23. Effect of $CaCO_3$ increments on Fe, Mn, and Zn extracted by different methods.[1]

Added $CaCO_3$	NH_4OAC Ext[2]			Easily Red.			Na_2EDTA Ext.		
	Fe	Mn	Zn	Fe	Mn	Zn[3]	Fe	Mn	Zn
				μg/l g Soil					
0	2.1	15.1	2.4	3.7	125.8	—	121.7	231.4	15.9
5	1.8	5.3	2.2	3.4	131.9	—	59.3	78.4	10.7
10	1.5	3.8	1.9	3.3	123.7	—	49.7	55.4	8.1
20	1.4	2.9	1.7	2.3	115.4	—	31.7	19.2	5.6
40	1.3	2.8	1.3	1.9	90.6	—	29.9	18.1	4.4

[1] Average values of 4 alluvial soils originally containing 3–6% $CaCO_3$.
[2] NH_4OAC used for Fe and Mn extraction had pH 7, while that used for Zn had pH 4.6.
[3] Not determined.
Source: R. El-Khomsy,[62] M.Sc. Thesis.

Table 24. Effect of Na_2CO_3 on Fe and Mn uptake by maize and their ratio in root:shoots (R:S).

	Fe		Mn	
	R:S	μg/Plant	R:S	μg/Plant
Untreated plants	0.62	188	0.23	16.3
Na_2CO_3	2.10	112	0.39	13.8

Source: Balba and El Khatib.[41]

Phosphorus-Micronutrients Relationships

Because of the ability of the calcareous soils to "fix" the applied soluble phosphorus, it was suggested that phosphate be applied at relatively high rates to satisfy the P "fixation" capacity of the soil. Several workers, however, had shown that micronutrients especially Fe availability to the plants supplied with excessive amounts of P is decreased.

Work by Alwash[43] showed that application of increments of P up to 6 mg/100 g soil successively decreased the NH_4OAC^- and EDTA-Fe extractable to about half the extracted amount without P. Also, the high application of P induced lower Fe uptake by wheat seedlings. Recognizing that plants growing on $CaCO_3$-rich soils are low in Fe content, any further decrease in Fe absorption—due to P application—might cause Fe deficiency in the growing plants. The direct cause of lower Fe absorption in the presence of high P applications is not clear. This phenomenon might be attributed to any or all the following factors:

a. Decrease of the available Fe amount.
b. Competition between Ca of the applied phosphate with the cationic Fe.
c. Competition between PO_4 with the anionic chelated Fe.

Table 25.   Fe concentration and percent Fe derived from applied ^{59}Fe in plant parts as affected by increments of $CaCO_3$.

CaCO$_3$%	^{59}Fe-Applied mg kg^{-1}	Fe Conc. $\mu g\ g^{-1}$		Fe % from Applied	
		R	Sh	R	Sh
0.8	0	89.2	5.7	—	—
0.8	50	103.9	6.1	2.3	7.2
0.8	100	111.8	6.3	4.7	8.7
0.8	150	108.7	6.4	4.9	13.2
10	0	12.8	4.2	10.2	—
10	50	44.7	4.1	1.6	2.4
10	100	77.4	5.9	3.8	7.8
10	150	87.2	4.3	4.3	7.3
20	—	10.4	3.8	—	—
20	50	45.1	4.2	1.5	2.7
20	100	78.8	4.5	2.2	2.6
20	150	78.3	4.8	3.0	4.1

R = roots; Sh = shoots.
Source: Balba and Osman.[65]

As for the effect of phosphorus application on Zn uptake by plants, Shehata[66] results showed that the amount of Zn absorbed by corn plants was not significantly affected by P increments. Contradicting results were found in the literature. Ellis et al.[67] showed that Zn concentration in beans and maize decreased upon application of 437–655 lbs P/acre. Takkar et al.[68] showed that severe Zn deficiency symptoms appeared on corn plants when P was applied at high rates. Similar results were reported by Stukenholtz et al.[69] and Bingham et al.[70] On the other hand, Racz et al.[71] found that Zn concentrations in plant tops were usually higher when 10 and 100 ppm P were applied than with 2 ppm P. Halim et al.[72] found that high P application did not appear to reduce the uptake of zinc by corn plants.

No direct study of the effect of P on Mn uptake was carried out by this author and his group, but a study had shown that with the application of $MnSO_4$ the concentration of P in the corn plant parts tended to decrease. The ratio of P in the roots : P in the shoots increased from 0.166 in plants grown without Mn application to 0.283 when $MnSO_4$ was added to the soil (Balba and El Khatib[41]).

Tentative deficiency levels have been established of 4.5 ppm for Fe, 1.0 ppm for Mn, 0.5 ppm for Zn, and 0.2 ppm for Cu (Fuehring[52]). Zinc symptoms were described by Halim et al.[72]

Foliar sprays are usually practiced on fruit trees. Six to eight kg of Zn per hectare broadcast or incorporated in the soil are usually sufficient. Because Zn deficiency is pronounced in maize, Zn is applied to correct its deficiency. Rate of Zn application to correct its deficiency in maize on highly calcareous soils may reach as high as 90 kg/ha (Fuehring[52]). Use of chelated Zn reduces the amount of Zn needed to between one-half and one-fifth.

Mn and Cu are much less apt to be deficient on calcareous soils than iron and zinc. If deficient, the sulphate forms are commonly applied to the soil.

GYPSIC SOILS

In arid and semiarid ecosystems, soil formation processes or conditions may lead to soil rich in gypsum. Gypsum is found in these soils as accumulated layers, veins, lenses, or diffuse amounts in the solum. It may be present in low insignificant amounts or in considerable percentages.

Gypsum Forms

The stable salt of hydrated calcium sulfate at ordinary temperatures is gypsum $CaSO_4 2H_2O$. It is present in nature in different forms: gypsum rock, gypsite, alabaster (a massive fine grained translucent variety, softer and lighter variety of gypsum) satinspar of fibrous silk form, and selenite transparent cleavable masses.

Crystals of gypsum are clear, white to grey yellowish and brownish in color. It may be tinted red, colorless, blue, etc., due to impurities (Twenhofel[73]).

Gypsum in Soil

Gypsum is found in soils of the arid and semiarid ecosystems in amounts ranging from traces to high percentages. In some soils, gypsum is present in sedimentary deposits from which the soil was derived, whereas in others it is formed by precipitation of calcium and sulfate during salinization. Gypsum formation processes vary from one place to another according to different environmental conditions as temperature, pressure, freezing, concentration of the media, organic matter, and microorganisms. When the groundwater is brought upward by capillary action to replace water lost by evaporation, the substances in the soil solution are deposited on or just beneath the soil surface. The deposits consist of everything carried in solution but $CaCO_3$ is generally the most abundant. Calcium, sodium, and potassium sulfates are also common (Twenhofel[73]).

Van Alphen and Romero[74] noted that gypsum is redistributed in the soil and frequently forms cemented and indurated layers. These layers form a mechanical impediment to root growth and have adverse properties of water retention and movement.

Under natural conditions $CaCO_3$, due to its lower solubility, is precipitated at shallow depths. Gypsum, being more mobile is either precipitated below the $CaCO_3$ horizon or leached off the solum depending on the depth of wetting following rainfall or irrigation. Also, the evaporites of lakes contribute to the gypsum formation. The composition of the lakes deposits is related to the rocks of the origin in which the lakes are situated. In complex solutions such as sea water, anhydrite is commonly the first formed precipitate. It may remain unchanged or it may be hydrated and transformed to gypsum. In other deposits gypsum is first precipitated. It may remain unaltered, but most likely it is to be converted to anhydrite either by reaction with concentrated brines or by pressure (McDonald[75]).

Theories of Gypsum Formation in Soils

Among the soil formation processes, oxidation-reduction have pronounced effect upon the composition, chemical activity, color, biological population, and other properties of recent sediments. Deposition of hematite indicates aerated conditions, while iron sulfides (pyrite or mascarite) signify reducing oxygen deficient medium. Saderite is an indication of intermediate oxidation potential.

It is believed that both oxidation and reduction processes can take place in the deep groundwater zones. At oxidation, the oxygen is carried downward by moving waters, then the sulfides change to sulfates and the pH is acidic. At reduction, the oxygen has been used up due to oxidation of organic substances, then sulphates are reduced and the pH is alkaline.

Role of Microbial Activity in Forming Sulfides

Bacteria influence the formation of iron sulfides as follows (Twenhofel[73]):

- Decomposition of sulfur-bearing proteins forming hydrogen sulfide.
- Sulfides of any origin may react with CO_2 and water to form hydrogen sulfide.
- Certain bacteria may act directly on free sulfur to form H_2S.
- SO_4-reducing bacteria, such as *Vibrio hydrosulfurous* and *Bacterium hydrosulfureum ponticum*, in the presence of decaying organic matter take oxygen from sulfates, thiosulfates, and sulfites to form sulfides.
- Sulfurous compounds may be oxidized with participation of sulfobacteria forming sulfuric acid as follows, Rozanov[76]:

$$FeS_2 + H_2O \rightarrow Fe(OH)_2 + S_2$$
$$2Fe(OH)_2 + S + 2H_2O + 2O_2 \rightarrow 2Fe(OH)_3 + H_2SO_4$$
$$H_2SO_4 + CaCO_3 + H_2O \rightarrow CaSO_4 + CO_2$$

Chemical Reactions

The reaction which occurs between calcium bicarbonate and magnesium sulfate solution produces gypsum and $Mg(HCO_3)_2$.

Pedological Classification

The gypsic soils belong to the order Aridisols, suborder Orthids, and the great group Gypsiorthids (Soil Survey Staff[77]).

The Typic Gypsiorthids are characterized with the following:

a. They do not have a petrogypsic horizon whose upper boundary is within 1 m of the soil surface.
b. They have a gypsic horizon in which the product of the percentage of gypsum and the thickness in centimeters above a depth of 1.5 m is 3000 or more.

Calcic Gypsiorthids are like Typic Gypsiorthids except for "b" and have a calcic horizon above the gypsic horizon.

Cambic Gypsiorthids are like Typic Gypsiorthids except for "b" and do not have a calcic horizon above the gypsic horizon.

Petrogypsic Gypsiorthids are like Typic Gypsiorthids except for "a" with or without "b."

The Gypsic Soils in the FAO/UNESCO Soil Classification[78]

The gypsic soils were included in an independent soil unit according to the presence of a gypsic horizon. This unit was divided into the following subunits:

- Typic Gypsisols contains a gypsic horizon.
- Petric Gypsisols contains indurated gypsic horizon.
- Calcic Gypsisols contains a calcic horizon.
- Luvic Gypsisols contains a clay horizon.
- Haplo Gypsisols other gypsic soils.

Effect of Gypsum on Soil Properties

As a soil constituent, gypsum plays important roles in the soil physical and chemical properties. These roles are reflected on the soil as a medium for plant growth.

Gypsum is directly added to soils suffering from sodicity and indirectly to normal soils with the superphosphate which contains about 50% of gypsum.

The physical effect of gypsum on the soil is related to its calcium as it acts as a coagulating agent when applied to fine-textured exchangeable sodium-rich soils. It is also added to the soil to counteract sodium-rich waters. Because gypsum is relatively soluble, its presence in relatively high concentrations in the soil system may initiate mechanical problems:

- The Ca^{++} ions usually compete with K^+ and NH^+_4 for plant absorption. Also Ca^{++} may accentuate PO_4 precipitation, thus decreasing P mobility to the plant root zone.
- Soils containing gypsum may create problems when irrigation is introduced. Water infiltrated from newly dug canals dissolve gypsum in the canal banks. The banks may collapse. Lining the canal banks is a temporary solution as the collapse may start upon irrigation of the land adjacent to the canal and extends to the canal banks.

Effect of Gypsum on Plant Growth

Most of the studies carried out on the gypsum effect on the growth of plants were in salt-affected soils. Loomis[79] studied the effect of heavy gypsum application. He found no significant effect on the growth of maize or soybeans, but the yield of oats had increased. Moreno and Osborn[80] found that gypsum application enhanced the uptake of several nutrients by plants like Ca, Mg, N, P, K, and S in sodic soils. Rasmussen et al.[81] found that application of 12 to 20 tons per acre $CaSO_4$ to seabree slick spots soils resulted in a seven-fold increase in yields of wheat and lucern.

Van Aiphen and Romero[74] concluded that up to 2% gypsum in the soil favors crop growth, that between 2 and 5% has little or no adverse effect if in powdery form, but more than 25% can cause a substantial reduction in crop yield. They attributed these yield reductions in part to imbalanced ion ratios with particular reference to K/Ca and Mg/Ca ratios.

Kandil[64] moistened and dried soil-gypsum mixtures several times before being planted to wheat or faba beans in pots. After 60 and 70 days of growth, respectively, the soil of each pot was analyzed for soluble cations and anions, exchangeable cations, P, Fe, and Mn.

The following results were reported:

- The soluble Ca, Mg, and K increased.
- The exchangeable Ca increased at the expense of exchangeable Mg, K, and Na.
- The $NaHCO_3$-extractable P was not affected.
- The NaOAC-extractable Fe and Mn were slightly increased.

The changes in Ca, Mg, Na, K, and P in the plants were slight.

MANAGEMENT OF CALCAREOUS SOILS

Lack of water is the main problem in arid ecosystems whatever the soil characteristics might be. With this prerequisite requirement satisfied, the $CaCO_3$-rich soils' utilization is controlled by the effects of the $CaCO_3$ content on the soil physical and chemical character- istics. Among these characteristics which are affected by $CaCO_3$ are:

- Soil-water relations and constants, presence of hardpans, infiltration of water, crustation, tilling practices, and machinery. Also reaction of soil $CaCO_3$ with applied nutrients plays an important role in fertilization and productivity level.
- $CaCO_3$-rich soils usually prevail in arid ecosystems while under humid ecosys- tems, $CaCO_3$ is usually leached to the bottom of the soil profile and hardly constitutes any problem.

From the above-described physical and chemical characteristics of these soils, the following management practices are recommended.

Land Leveling

It has been noticed that shallow profiles are frequently encountered in regions of calcareous soils. Leveling these soils to be as flat as for flood irrigation is not recommended in areas with such shallow profiles. This practice may end up with areas with shallower profiles than before leveling. The uncovered subsoil is usually less fertile than the upper surface layer. If leveling is carried out, a survey for the profile depth of the area has to be made. The land use pattern may have to be changed due to changes in the profile depth after leveling.

Irrigation Method

The selection of the suitable irrigation method depends on several factors. However, since fine leveling is not recommended when they are characterized by undulated surface such as the calcareous soils in the Western Desert of Egypt, the contour irrigation as well as the sprinkler methods are more suitable than the flood irrigation method. Drip irrigation may also be utilized.

From a practical standpoint Massoud[6] recommended that in order to have good wheat stand in calcareous soils, the moisture tension of the crust during emergence should be kept below about 0.33 atm and that the planting depth be relatively shallow, less than 4 cm. The amount of seeds needed for planting may also be increased.

Soil Tillage

The study carried out by Fuller and Padgett[15] showed evidence which supports the idea that intensive tilling is an important factor in the deterioration of a favorable physical condition in the soil. The more intensive the tilling, the greater was the tendency of the soil structure to break down and the soil to form crusts.

Because these soils change their favorable properties when irrigated, they become indurated and resistant to root penetration when they are dry, especially in that portion of the profile subjected to wetting and drying. Therefore, the depth of plowing is one of the important factors in relation to the success or failure of growing crops.

Several studies were conducted on different plowing technique and implements. The selection of the right plough type, tillage sequence, plowing depth, and moisture content at the time of plowing should provide good soil tilth. Fuller and Padgett[15] compared the influence of three methods of tillage: fallow (undisturbed), discing, and rototilling, on the aggregation of Jokake clay loam and was compared both with and without soil conditioner. Their results show that the percent aggregates in the two size groups greater than 0.015 mm and less than 0.05 mm were greatest for the fallow. More aggregation was found in disced than rototilled soil. The difference was significant at only a few samplings during the two-year period. The results of Massoud et al.[82] indicated that the optimum depth of plowing should not be less than 20 cm, and preferably 25 cm using a mould board plough followed by a chisel plough in a perpendicular direction. The range of moisture content suitable for plowing is very narrow and occurs 4–5 days after irrigation. After 7–8 days, plowing becomes difficult.

The problem of crusting has been pointed out above. In this regard, Jackson and Erie[83] reported that bitumen may be used as a treatment to reduce soil crusting. They stated that bitumen may be sprayed on a small area directly over the seed after planting, or as a band over the seed row. In addition to reducing crusting, the black material increases the absorption of solar radiation and therefore soil temperature. It also reduces evaporation of water. These factors improve the seed environment. They added that experiments have shown the feasibility of this treatment, but it has not been generally adopted as a management practice.

In soils stabilized with VAMA soil conditioner, Fuller and Padgett[15] reported that the difference in aggregation was significant between disced and rototilled plots. On the other hand, untilled plots having VAMA conditioner were very often found to have significantly greater aggregation than disced or rototilled plots. The soil receiving VAMA was always better aggregated than that not receiving VAMA despite the intensive tilling practice. In general, these findings are also evident in soils receiving IBMA, which stabilized soil aggregates somewhat better than HPAN conditioner.

Irrigated calcareous soils are susceptible to compaction by heavy equipment. Harris et al.[84] found that compaction by the tractor wheels on a plowed calcareous soil had decreased infiltration rates in the compacted areas to 43% of those in noncompacted areas.

Impermeable Layers

In areas of calcareous soil, soil profiles containing hard pans or relatively impermeable layers are frequently encountered. Presence of such pans hinders water movement, decreases the soil air proportion in the soil pores and limits growth of plant roots. Under certain conditions a perched water table may be formed and secondary salinization of the soil occurs.

Soils containing such hard pans should be deep-plowed, with chisel, deep moldboard plow to depths of 1.2 or 1.5 m. Large powerful machinery is needed for this operation and must be done when the soil is relatively dry in order to break up the impervious soil layers (Jackson and Erie[83]). After destructing these pans, efficient drainage should be established.

In case of relatively deep impermeable layers, deep plowing might be an expensive practice. In this case, it might be advisable to utilize the soil with its shallow profile. The depth of the drains should be limited by the depth of the soil profile. The drain spacing should be narrowed. Irrigation water should be calculated to moisten such a limited depth. Plants with surface roots are preferred and fertilization should be increased.

Drainage

Drainage is as important for such soils as for others. The field drains should not be cut in the impermeable parent material, rather, their depth is limited by the depth of the profile.

Fertilization

The calcareous soils are generally poor in nitrogen and phosphorus. Their low content of both elements is the main cause of their low productivity.

A plan to supply these soils with organic materials is recommended. Beside its content of nutrients, organic matter helps to improve the physical properties of these soils. If a legume crop can be turned under, it improves the soil structure, decreases crust formation, and supplies the growing crops with its content of nutrients after its decomposition.

The main points of N and P reactions with soil upon their application were discussed above. Potassium is generally present in adequate amounts, yet crops known to need much K such as potatoes require its application.

It is recommended that fertilization be based on soil and plant-tissue testing.

The common soil test for available N is the determination of nitrates formed after incubating the soil for several days (Munson and Stanford[85]).

Smith and Stanford[86] suggested a method which involves autoclaving the soil for 16 h (Fuehring[52]). The nitrate electrode method makes the determination of nitrates possible.

Nitrogen fertilizers may be applied any time from just before planting up to the time the plant is well established. They are biologically transformed into nitrate within a period of few days. The nitrate is subject to loss by leaching with irrigation or rain water. Surface application of NH_4-N carriers or N carriers that form NH_4 in the soil as the urea should be avoided. The loss of NH_3 from NH_4 fertilizers is greatly enhanced when placed on the $CaCO_3$-rich soil surface (Terman and Hunt[24]).

The rates of N application for most crops are 1 to 1.5 times the expected amount of N removal in the crop less than the amount estimated to be available in the soil (Fuehring[52]).

The virgin $CaCO_3$-rich soils are known for their low P content. Soluble phosphates applied to these soils are subject to adsorption and precipitation reactions. However, a proportion of the applied soluble phosphate remains available to plants, Balba et al.[87]

Olsen and Flowerday[89] recommended that the applied fertilizer should be at least 20% water soluble.

Because recovery of applied P is usually less than 20% during a single growing season, there is considerable residual effect (Campbell[90]). Thus, calcareous soils which tested poor will test rich in available P after several seasonally applications (Balba[88]).

Phosphorus is broadcast followed by plowing before seeding. Band application of phosphates proved more effective than broadcasting (Balba[88]). Applied amounts of phosphatic fertilizers depend on soil testing, and cost of the fertilizer. From field experiments carried out in each location, the optimum P application may be calculated (Balba[91]).

Several methods for soil testing for micronutrients have been introduced. The method of Lindsay and Norvell[92] using the DTPA extract is widely acceptable and utilized in many laboratories. Deficiency levels of 4.5 mg kg^{-1} for Fe, 1.0 mg kg^{-1} for Mn, 0.5 mg kg^{-1} for Zn, and 0.2 mg kg^{-1} for Cu have been tentatively established (Fuehring[52]).

Iron and zinc deficiency symptoms are more common on plants grown on calcareous soils than Mn or Cu deficiencies. Plant tissue analysis together with soil testing should ascertain deficiency symptoms before recommending their application.

The amount of applied soluble nutrients, especially N, should be divided into more than one application. Fertilizers added with or close to the seed may impede or prevent germination due to the salt effect.

These soils are relatively poor in micronutrients. Application of Fe, Zn, and Mn soluble salts to the soil may not solve the problem of their deficiency since they might be immobilized in the plant roots or precipitated in the soil. The chelated forms of nutrients are recommended.

Spraying the nutrients on the plant leaves seems to be the most efficient method to supply the plants growing on calcareous soils with their need of macro- and micronutrients.

Selection of Crops

When these soils are cropped the first time, one should take into consideration:

a. Depth of profile.
b. Amount of available water.
c. Method of irrigation.

The above-stated factors are those related to a specified area. Other factors such as labor, mechanization, marketing, etc., apply to the selection of any cropping system in general and have to be taken into consideration.

Areas with shallow profiles might be left as pasture land or planted to forage crops. Orchards are to be located in the areas with deep profiles.

The most successful crops are those grown locally in the area. Pilot areas to test kinds and varieties of plants are helpful in selecting the most suitable crops under the local environments.

Calcium-loving plants such as legumes are successful in areas of calcareous soils. Several vegetables, especially cuckerbitae such as squash and melons and solanaceae as tomatoes, peppers, and eggplants, proved successful in the calcareous soils.

In the calcareous soils of Egypt, under the Mediterranean climate, olives, figs, grapes, and dates have been growing since ages ago.

The amount of available water, if limited, will limit the selection of crops to those with low water consumptive use.

If drip irrigation has to be utilized to conserve as much water as possible, only vegetables or orchards can be grown.

Plants differ in their ability to tolerate $CaCO_3$. Even within one kind of plant, varieties differ in their ability to resist the calcareous conditions. This ability should be taken into consideration.

REFERENCES

1. FAO/UNDP Symposium on Calcareous Soils, Cairo, FAO Soil Bull., 21:300, 1973.
2. Reullan, A., Morphology and distribution of calcareous soils in the Mediterranean and desert regions, FAO Soil Bull., 21:7, 1973.

3. de Meestere, T., Morphology, mechanical composition and formation of highly calcareous lacustrine soils of Turkey. FAO Symp. on Calc. Soils Bull., Cairo, 1973.

4. Buehrer, T.F. and J.A. Williams, The hydrolysis of calcium carbonate and its relation to the alkalinity of calcareous soils. University of Arizone Tech. Bull., 64:41, 1936.

5. Menon, R.C., Report of the work done in the Soil Laboratory, UNSF Ghab Development Project, Syria, 1971. Cited in Massoud.[6]

6. Massoud, F.I., Some physical properties for calcareous soils and their related management practices. FAO Soil Bull. 21:73, 1973.

7. Massoud, F.I., M.M. El Gabaly, and E. Talty, Moisture characteristics of the highly calcareous soils of Mariut Extension Project, Egypt, Alex. J. Agric. Res. 19:351, 1971.

8. El Gabaly, M.M., Reclamation and management of the calcareous soils of Egypt. FAO Soil Bull. 21:123, 1973.

9. Balba, A.M. and M.F. Soliman, Salinization of homogeneous and layered soil columns due to upward movement of saline ground water and its evaporation. Agrochemica, 13:542, 1969.

10. Childs, E.C. and N. Collis-George, The permeability of porous materials. Proc. Royal Soc. 201:A 392, 1950, Cited in Massoud.[6]

11. Hillel, D., Introduction to Soil Physics. Part II. The Solid Phase, p. 51. Academic Press, New York, 1982.

12. Richards, L.A., Modulus of rupture of soils as an index of crusting of soils, Soil Sci. Soc. Am. Proc., 17:321, 1953.

13. Allison, L.E., Soil and plant response to VAMA and HPAN soil conditioners in the presence of high exchangeable sodium. Soil Sci. Soc. Am. Proc., 20:147, 1956.

14. Lemos, P. and J.F. Lutz, Soil crusting and some factors affecting it. Soil Sci. Soc. Am. Proc., 21:485, 1957.

15. Fuller, W.H. and C.G. Padgett, The effect of discing, rototilling and water action on the structure of some calcareous soils. University of Arizona Tech. Bull., 134:26, 1958.

16. Nakayama, F.S., theoretical consideration of the calcium sulfate-bicarbonate-carbonate interrelations in soil solutions. Soil. Sci. Soc. Am. Proc., 33:698, 1969.

17. Doner, H.E. and P.F. Pratt, Solubility of $CaCO_3$ precipitated in aqueous solutions of magnesium and sulfate salts. Soil Sci. Soc. Am. Proc., 33:690, 1973.

18. Seatz, L.F. and H.B. Peterson, Acid, Alkaline, Saline and Sodic Soils, Chapter 7. in the Chemistry of the Soil. F. Bear (Ed.) 2nd ed. A.C.S. Monogram No. 160, 1965.

19. Amrhein, C., J.J. Jurinak, and W.M. Moore, Kinetics of calcite dissolution as affected by CO_2 partial pressure. Soil Sci. Soc. Am. J., 49:1393, 1985.

20. Clark, J.S., An examination of the pH of calcareous soils. Soil Sci., 98:145, 1964.

21. Balba, A.M., Effect of sodium water and gypsum increments on soil chemical properties and plant growth. Alex. J. Agric. Res., 8:51, 1960.

22. US Sal. Lab. (Richards, Ed.), Diagnosis and Improvement of Saline and Alkali Soils. USDA Handbook No. 60, 1954.

23. Balba, A.M., M.G. Nasseem, and S. El Massry, Soil fertility studies of the N.W. Coast of UAR. I-Factors affecting utilization of nitrogen. J. Soil Sci. UAR, 9:25, 1969.

24. Terman, G.L. and D.M. Hunt, Volatilization losses of nitrogen from surface applied fertilizers as measured by crop response. Soil Sci. Soc. Am. Proc., 28:667, 1964.

25. Balba, A.M. and M.G. Nasseem, The loss of ammonia by volatilization from nitrogenous fertilizers added to the soils. J. Inst. Trop. Agric., 3:213, 1968.

26. Fenn, L.B. and D.E. Kissel, Ammonia volatilization from surface applications of ammonium compounds on calcareous soils. I-General theory. Soil Sci. Soc. Am. Proc., 37:855, 1973.

27. Fenn, L.B. and D.E. Kissel, Ammonia volatilization from surface applications of ammonium compounds on calcareous soils. II- Effects of temperature and rate of NH4-N application. Soil Sci. Soc. Am. Proc., 38:606, 1974.

28. Stumpe, J.M., P.L.G. Velk, and W.L. Lindsay, Ammonia volatilization from urea and ureaphosphates in calcareous soils. Soil Sci. Soc. Am. J., 48:921, 1984.
29. Solis, P. and J. Torrents, Phosphate fractions in calcareous vertisols and inceptisols of Spain. Soil Sci. Soc. Am. J., 52:462, 1989.
30. Boischot, M.C. and J. Herbert. Fixation de l'acide phosphorique sur le calcium de sols. Plant and Soil, 2:311, 1950.
31. Cole, C.V., S.R. Olsen, and C.O. Scott, The nature of phosphate sorption by $CaCO_3$. Soil Sci. Soc. Am. Proc. 17:852, 1953.
32. Balba, A.M., The phosphates formed on reactions between calcium hydroxide and orthophosphoric acid. UAR. J. Soil Sci., 4:439, 1964.
33. Olsen, S.R., C.T. Cole, F.S. Watanabe, and L. Zean, Estimation of available phosphorus in soils by extraction with sodium bicarbonate. Circ. No. 939, USDA, Gov. Printing Office, Washington, D.C., 1954.
34. Beckette, P.H.T. and R.E. White, Studies on the phosphate potentials of soils: 3—The pool of labile inorganic phosphate. Plant and Soil, 21:253, 1970. Cited in Olsen,[21] 1973.
35. Jensen, H.W., Phosphate potential and phosphate capacity of soils. Plant and Soil, 33:17, 1970.
36. Olsen, S.R., Nutrient supply and availability in calcareous soils. FAO Soil Bull., 21:41, 1973.
37. Balba, A.M., Potassium forms and sufficiency to plants in Egyptian soils. J. Inst. Trop. Agric., 5:19, 1965.
38. Abdel Kader, F. and S.I. Abu Ghalwa, Distribution of total and free iron forms in different soils of Egypt. Alex. J. Agric. Res., 2:443, 1973.
39. Kishk, F. and I. Mohammed, Zinc in calcareous soils. Plant and Soil., 39:497, 1973.
40. Ghanem, I., M.N. Hassan, M. Xhadr, and V.T. Tadros, Studies on manganese in soils. I-Status of Mn in some selected soils of Egypt. J. Soil Sci. UAR, 11:113, 1971.
41. Balba, A.M. and S. El Khatib, Available soil Mn and its distribution in corn plants under variable soils conditions. Alex. Sci. Exch., 1:53, 1980.
42. Thabet, A.Y.G., Iron relationships with soils and plants. M.Sc. Thesis, University of Alexandria, Egypt, 1976.
43. Alwash, H.K., Phosphorus and Iron Requirements of High Yielding Varieties of Wheat. M.Sc. Thesis, University of Alexandria, Egypt, 1975.
44. Halverson, H.O. and I. Starkey, Studies on the transformation of iron in nature. Soil Sci., 24:3, 1937.
45. McGeorge, W.T., Studies on plant food availability in alkaline calcareous soils. Ariz. Agric. Exp. Sta. Tech. Bull. 94, 1942.
46. Inskeep, W.P. and P.R. Bloom, Soil chemical factors associated with soybean chlorosis in Calcioquolls of Western Minnesota. Agron. J., 79:779, 1987.
47. Bureau, A.G., An investigation of soil factors in iron deficiency chlorosis of soybeans. Ph.D. Diss. University of Minnesota, St. Paul, 1963. Cited in Inskeep.[45]
48. Loeppert, R.H., L.R. Hossner and M.H. Chmeilewski, Indigenous soil properties influencing the availability of iron in calcareous hot spots. J. Plant Nutr., 7:135, 1984.
49. Brown, J.C., O.R. Lunt, R.S. Holmes, and L.O. Tiffin, The bicarbonate ion as an indirect cause of Fe chlorosis. Soil Sci., 88:260–266, 1959.
50. Kadry, L.T., Distribution of calcareous soils in the Near East region, their reclamation and land use measures and achievements. FAO Reg. Sem. on Calc. Soils, Bull., 21, 1973.
51. Hartwig, R.C. and R.H. Loeppert, Pretreatment effect on dispersion of carbonates in calcareous soils. Soil Sci. Soc. Am. J., 55:19, 1991.
52. Fuehring, H.D., Response of crops grown on calcareous soils to fertilizers. FAO Soil Bull., 21:53, 1973.
53. Miller, G.W. and D.W. Thorne, Effect of biocarbonate ion on the respiration of existed roots. Plant Phys., 31:151.

54. Miller, G.W., J.C. Brown, and R.S. Holmes, Chlorosis in soybean as related to iron, phosphorus, bicarbonate and cytochrome oxidase activity. Plant Phys., 35:619, 1960. Cited in Brown and Jolley,[31] 1986.
55. Coulomb, B.A., R.L. Chaney, and J. Wiebold, Bicarbonate directly induces iron chlorosis in susceptible soy bean cultivars. Soil Sci. Soc. Am. J., 48:1297, 1984.
56. Brown, J.C. and V.D. Jolley, An evaluation of concepts related to iron-deficient chlorosis. J. Plant Nutn., 9(3–7):175, 1986.
57. Mengel, K., W. Bull, and H.W. Scherer, Iron distribution in vine leaves with bicarbonate-induced chlorosis. J. Plant Nutr., 7:715, 1984.
58. Loeppert, R.H., Reactions of iron and carbonates in calcareous soils. J. Plant Nutr., 9(3–7):195, 1986.
59. Lindsay, W.L. and D.W. Thorne, Bicarbonate ion and oxygen level as related to chlorosis. Soil Sci., 77:271, 1954. Cited in Loeppert,[33] 1986, p. 201.
60. Miller, G.W. and H.J. Evans, Inhibition of plant cytochrome oxidase by bicarbonate. Nature 178:974, 1956. Cited in Loeppert,[33] 1986.
61. Romheld, V., H. Marschner, and D. Kramer, Response to Fe deficiency in roots of Fe efficient plant species. J. Plant Nutr. 5:489, 1982. Cited in Loeppert,[33] 1986, 201.
62. El-Khomsy, R., A Study of the Effect of Some Factors on the Soil Available Forms of Iron, Mn and Zn. M.Sc. Thesis. University of Alexandria, Egypt, 1977.
63. Balba, A.M., A. Osman, and I. Ghattas, Iron distribution in faba bean plants irrigated with NaHCO₃-water. Zeit Schrift Pflan. und Boden Runde, V., 143:268, 1980.
64. Kandil, K., A study of gypsum in soils and its effect on plant. M.Sc. Thesis, University of Alexandria, Egypt, 1978.
65. Balba, A.M. and A. Osman, Final Report on Project No. 28 A. Mid. East Center Rad. Isotope, Cairo, 1980.
66. Shehata, N.Y., Response of Corn and Soybean Plants to Different N Fertilizers and CO₂ Enrichment. Ph.D. Thesis, Catholic University of Louvain, Belgium.
67. Ellis, R., J.V. Davis, and D.L. Thurlovo, Zinc availability in calcareous Michigan soils as influenced by phosphorus level and temperature. Soil Soc. Am. Proc., 28:83, 1969.
68. Tuckar, P.N., M.S. Mann, N.S. Randhawa, and S. Hardev, Yield and uptake response of corn to zinc as influenced by phosphorus fertilization. Agron. J., 68:941, 1976.
69. Stukenkoltz, D.D., R.J. Olsen, G. Gogen, and R.A. Olson, On the mechanism of phosphorus-zinc interaction in corn nutrition. Soil Sci. Soc. Am. Proc., 30:759, 1966.
70. Bingham, F.T., J.P. Martin, and J.A. Chastain, Effects of phosphorus fertilization of California soils on minor element nutrition of citrus. Soil Sci., 86:24, 1958.
71. Racz, G.J. and P.W. Haluschak, Effects of phosphorus concentration on Cu, Zn, Fe and Mn utilization by wheat. Ca. J. Soil Sci., 54:357, 1974.
72. Halim, A.H., C.E. Wassom, and R. Ellis, Zinc deficiency symptoms and zinc phosphorus interactions in several strains of corn. Agron. J., 60:267.
73. Twenhofel, W.H., Principles of Sedimentation. McGraw-Hill Book Co., New York, 1950. Cited in Kandil.[79]
74. Van Alphen, J.G. and F. de Los Rios Romero, Gypsiferous Soils. Int. Inst. Land Recl. and Improvement. Bull. 12, 1971.
75. McDonald, G., Anhydrite-gypsum retention. M. J. Sci., 251:884, 1953, Cited in Kandil, 1979.
76. Rozanov, B.C., The serozem of Central Asia. Office of Tech. Services, USDA Commerce, Washington, D.C., 25, Cited in Kandil.[64]
77. Soil Survey Staff, Soil Taxonomy, USDA Handbook No. 436, Washington, D.C., 1975.
78. FAO / UNESCO, Soil Map of the World. System, 1:5 M, UNESCO, Paris, 1974.
79. Loomis, W.E., Effect of gypsum applications on plant growth. Plant Phys., 19:706, 1944, Cited in Kandil.[79]

80. Moreno, E.C. and G. Osborn, Solubility of gypsum and dicalcium phosphate dihydrate in the system $CaOP_2O_5$-SO_4-H_2O and in salts. Soil Sci. Soc. Am. Proc., 27:614, 1963.

81. Rassmussen, W.W., G.C. Lewis, and M.A. Fosberg, Improvement of the Chilcott Sebree (sodolized-solonetz) slick spots soils in south western Idaho. (C.F. Soils and Fertilizers, 28:1288, 1965.) Cited in Kandil.[79]

82. Massoud, F.I., A.M. El-Hossary, M.M. El-Gabaly, and M.A. Mabrouk, Effect of different tillage implements and techniques on tilth of a highly calcareous sandy loam soil. Proc. UAR Soil Sci. Soc. Symp. on Calc. Soils., Alexandria, UAR, 1968.

83. Jackson, R.D. and J. Erie, Soil and water management practices for calcareous Soil. FAO Soil Bull. No. 21, 1973.

84. Harris, K., L.J. Erie, and W.H. Fuller, Minimum tillage in the southwest. University of Arizona Agr. Expt. Sta. Bull., A-39.

85. Munson, and G. Stanford, Predicting nitrogen fertilizer needs of Iowa soils. IV-Evaluation of nitrate production as a criterion of nitrogen availability. Soil Sci. Soc. Am. Proc., 19:461, 1955.

86. Smith, S.R. and G. Stanford, Evaluation of chemical index of soil nitrogen availability. Soil Sci., 111:228, 1971.

87. Balba, A.M., M.G. Nasseem, and N. Yuwakeem, Soil fertility of the N.W. Coast of UAR. II-Phosphorus and Potassium. J. Soil Sci. UAR, 9:85, 1968.

88. Balba, A.M., The relationship between plant and phosphorus in highly calcareous soils using ^{32}P-labeled superphosphate. J. Isot. and Rad. Res., 2:55, 1968.

89. Olsen, S.R. and A.D. Flowerday, Fertilizer phosphorus interactions in alkaline soils, 1971, in R.A. Olsen (Ed.) Fertilizer Tech. and Use, 2nd ed., Soil Sci. Soc. of Am. Madison, Wis., Cited in Fuehring.[51]

90. Campbell, R.E., Phosphorus fertilizer residual effects on irrigated rotations. Soil Sci. Soc. Am. Proc., 29:67, 1965, Cited in Fuehring.[51]

91. Balba, A.M., Quantifying Plant Relationship with Nutrients. Adv. Soil and Water Res. in Alexandria, No. 8–9, 1988–1989, Pub. A.M. Balba Group for Soil and Water Res., Alexandria.

92. Lindsay, W.L. and W.A. Norvell, Equilibrium relationships of Zn, Fe, Ca and H with EDTA and DTPA in soils. Soil Sci. Soc. Am. Proc., 33:62, 1969.

CHAPTER 4

Sandy Soils

INTRODUCTION

The sandy soils are soils with problems. The growers suffer from their unfavorable physical and fertility problems.

The coarse texture of these soils is responsible of their high infiltration rate of water. Thus plants growing on these soils usually suffer from drought unless frequently irrigated. The loss of water from the soil by infiltration is accompanied with loss of soil finer particles and of added soluble nutrients. The water is also lost by seepage during its conveyance to the fields.

These soils contain low organic matter and fine particles. Thus their cation exchange capacity is low. The exchangeable cations are considered available forms of nutrients to plants. Also, their nitrogen content is low due to their low organic matter. Because of their coarse texture the applied nutrients are subject to leaching. Because of the coarse texture and the single particles structure of these soils, they are subject to water as well as wind erosion.

Soils of Sandy Texture

The first of these groups are the sands which contain 85% or more of sand, with its content of silt plus 1.5 times the clay, do not exceed 15%.

The second group is the loamy sand which contains 85 to 90% sand as its higher limit, and silt plus clay not more than 15% as in the first group. Its lower limit content of sand is not less than 70 to 85% and a proportion of silt plus clay, provided the summation of silt content plus double the clay content does not exceed 30%.

Most of the constituents of the groups "sands" and "loamy sands" are made up of single particles not bound together, especially when the organic matter binding compound is low. Accordingly, the sandy soils have no structure.

FORMATION OF SANDY SOILS

The hot dry climate most of the year and the wind are the main environmental conditions in several regions of sandy soils. In winter, short rainy storms may take place. However, this rain barely moistens the surface, while the subsoil remains mostly dry. Gypsum and calcium carbonate, if present on the soil surface, may dissolve and penetrate through the soil to accumulate at a depth in the soil profile. The natural vegetative cover under such hot and dry conditions is sparse. Thus, the soil organic matter content is low and hardly is more than 0.2–0.5%.

The dryness, the sparse vegetative cover, and the low biological and chemical activities result in slow soil formation processes. In general, the sandy soils have a profile without distinguished horizons, especially in the regions having dry climates. In more humid regions, the profile may have a "b" clay horizon with a weak structure and a formation of clay minerals.

Gypsum and calcium carbonate are usually present in most sandy soils in arid and semiarid regions. A horizon of both materials is formed in the profile, especially in the semiarid regions. In the arid regions the calcium horizon is usually close to the soil surface.

A thin solid crust is usually formed on the sandy soils surfaces as a result of rainfall. This crust impedes the water infiltration into the soil depth and helps the formation of torrents and soil erosion by water. This crust may be made up of calcium as a result of accumulations of gypsum or calcium carbonate.

Sand transportation by wind is a common phenomenon in regions of sandy soils. In the wind-eroded areas, the gravel increases on the soil surface. The gravel and stone fragments on the soil surface have shining surfaces due to iron and manganese oxides. In the sandy soils regions, the sand dunes are frequent.

THE SANDY SOILS IN SOIL CLASSIFICATION SYSTEMS

The sandy soils fall in the "Entisols" and "Aridisols" orders of the U.S. Soil Classification (U.S. Soil Survey Staff[1]).

The Entisols are found in arid as well as humid regions, while the Aridisols prevail in arid regions.

The Entisols are soils either without natural genetic horizons or with only the beginning of horizons. Hard rock may be present at shallow depths. In arid regions, the Entisols may show small secondary accumulations of carbonates, sulfates, or more soluble salts, but not enough to constitute calcic, gypsic, or salic horizons. The sandy soils of the Entisols order fall mainly under the suborder Psaments. They include many of the soils formerly called Dry Sands and Very Sandy Alluvial Soils. They range in properties from calcareous sands on natural levees or recent dunes, to the quartz sand commonly found on coastal plains.

The Aridisols were described in the first chapter.

In the U.S. Soil Conservation Service, the sandy soils in the middle of United States are evaluated as being in the second and third degrees which can be cultivated with protection from erosion. Limited agricultural activities can be carried out in the soils of the fourth degree while soils of the fifth and sixth degrees are left for pasture.

In the soil survey for reclamation by using the High Dam water in Egypt, the sandy soils were evaluated as being in the fourth and fifth degrees.

In the Soil Map of the World[2] (SMW) by FAO/UNESCO, the sandy soils were classified as follows:

Order	**Entisols** Soils with no diagnostic horizons except an ochric horizon (SMW).
Suborder	**Psaments (FAO, Regosols)** Entisols with a texture of loamy fine sand or coarser between 24 and 100 cm depth.
Group	**Torripsaments** Psaments that are usually dry.

Suborder	**Fluvents (Fluvisols) (SMW)**
	Entisols in which organic matter decreases irregularly with depth or is less than 0.35% at 125 cm depth.
Group	**Torrifluvents**
	Fluvents that are usually dry.
Suborder	**Orthents (FAO, Regosols)**
	Other entisols.
Group	**Torriorthents**
Order	**Aridisols (FAO Xerosols)**
	Soils having an ochric epipedon that is not hard and massive and with one or more of the following:
	a- Dry with an argilic or natric, cambic, calcic, petrocalcic, duripan horizon, or
	b- EC 2 dS m^{-1}
	c- Saturated with water for 1 month or more and a salic horizon.
Suborder	**Orthids**
	Aridisols with no argilic or natric horizon.
Group	**Durorthids**
	Other orthids having a duripan.
	Salorthids
	Other orthids saturated with water for 1 month or more that have a salic horizon above any calcic or gypsic horizon.
	Paleorthids
	Other orthids with a petrocalcic horizon.
	Calciorthids
	Other orthids that are calcareous throughout and have either a calcic or gypsic horizon.
	Comborthids
	Other orthids

PHYSICAL CHARACTERISTICS

It has been stated above that the sandy soils are made up of particles having the size of sand. The relatively large size of particles results in large pores between the particles, though the total pores constitute low porosity. The porosity or pore space is that space between the particles which is equal to the ratio of the volume of voids either filled with air or with water to the total volume of soil, including air and water.

The apparent specific gravity or bulk density of the sandy soils is generally high, being 1.6–1.8 g/cm^3, while that of clay soils is 1.4–1.5 g/cm^3.

The drainable pore volume refers to the volume of water released or taken by a unit of volume of soil in the zone which may be under influence of a fluctuating water table. The sandy soils have a high content of large pores which results in larger channels for water conveyance. Thus, these soils have good drainage. Also, because water is usually retained in the narrow pores and not in the large ones little amounts of water may be retained in the sandy soils.

Because the sandy soils contain a higher proportion of air at the expense of their content of moisture, the air penetration through these soils decreases with the decrease in the size of particles.

The Soil-Water Relationships

The field capacity (FC) is defined as the percentage of water retained in the pore space of a soil after the excess water from an irrigation has percolated to deeper layers. In practice the field capacity is determined 1–2 days after an irrigation. Permanent wilting point (PWP) is defined as the percentage of water, still remaining in a soil, once the plants are no longer capable of extracting sufficient moisture to meet their needs.

One of the main characteristics of the sandy soils is their low moisture retention under low or high tension. Thus, these soils are sometimes described as being "droughty." Because of their low content of clay and the large size of pores, the majority of water that may applied to the sandy soils is lost by infiltration.

The range of available water to plants is the difference between the moisture percentages at the field capacity (FC) and the permanent wilting point (PWP). This range is limited to about 4–6%, while in the clay soils this range reaches about 16–29%. Evidently, this characteristic is of practical importance as it controls frequency of irrigation (Table 26).

The sandy soils retain water less tightly than the finer-textured soils. Thus, they give up water to the crop more readily than the fine-textured soils. This property, however, may be considered as an advantage or disadvantage for the sandy soils, depending on the kind of crop and other soil characteristics (USSWC Res.[3]):

- Crops grown for grain: If there is no water table and no fine-textured subsurface soil within the root zone, the crop will use up water faster than one grown on finer-textured soil. Also, it is subject to wilt before the next rain and no grain may be harvested.
- Crops grown for forage: No need to wait for the delayed rain. It could be harvested before the crop dries.

Infiltration Rate

The infiltration rate is the velocity of water movement within the soil. This rate slows when the time of contact between the soil particles and water increases until it reaches a constant value equal to the hydraulic conductivity. The rate of infiltration in the sandy soils is usually 2.5–25 cm/h or about 250 times the infiltration rate in the clay soils (0.01–0.1 cm/h). It may even increase in the sandy soils to 10–200 cm/h.

The loss of water from the sandy soils both in the conveyance systems and in the fields is related to the high infiltration rate. Surface flood irrigation is not recommended if the final infiltration rate reaches 10 cm/h or more. The fast rate of infiltration transports the fine particles from the soil surface to accumulate in the depth of the profile forming a layer of low permeability. A water table might be formed above such an impermeable layer.

In the case of a coarse-textured soil having a subsurface layer with fine texture, the infiltration rate at its start is fast. When the water reaches the depth between the coarse and the fine layers, the infiltration rate slows down until it matches the infiltration rate of the fine layer. But if the layer of fine particles on the surface is followed by a coarse-textured layer, the infiltration rate is, in this case, controlled by the fine-textured layer.

Moisture Distribution in the Soil

After water infiltration in the soil ceases, the distribution of soil moisture varies in the sandy soils at a faster rate and deeper in the soil than in the clay soils.

Table 26.   Field capacity, permanent wilting point (PWP) and available soil water.

Soil Texture	Water Content %		
	Field Capacity	PWP	Available
Coarse sand	8–10	3.5–4.5	4.5–5.5
Fine sandy loamy	14–17	6.0–7.5	8.0–9.5
Loam	20–17	7.5–9.5	9.5–10.5
Clay loam	19–24	9.5–11.0	9.5–13.0
Clay	27–35	15.0–19.0	12.0–16.0

Source: Land Master Plan.

The fast infiltration rate of water through the sandy soils may be considered advantageous or otherwise depending on several factors and conditions (USSWC Res.[3]):

- Because rain penetrates deeper in the sandy soils more than in fine-textured soils, less water is subject to loss by evaporation in sandy soils. Thus, in areas of low rain fall, a high infiltration rate can be considered beneficial.
- High infiltration rate in soils having a finer-textured soil at the lower part of the root zone may be advantageous after heavy rainfall. Water and nutrients are retained by this fine-textured subsurface soil for plant use.

If the soil is sandy to a considerable depth, water and nutrients may penetrate below the root zone depth and become unavailable to plants.

Area of the Specific Surface

The sandy soils are characterized with small surface areas for their particles compared with the surface area of the particles of the clay soils. Thus, reactions related to surfaces are weaker in sandy soils than those in clay soils.

In general, these soils are not elastic when moist and lose consistency when dry. Their apparent specific weight is relatively high (1.55–1.80 g/cm^3) and their total porosity is lower (32–42%) than in finer soils.

MANAGEMENT OF THE SANDY SOILS

The management practices of the sandy soils are controlled by their physical properties and their fertility status.

The physical properties cause losses of water from the soil by infiltration deeper than plant root zone and seepage from the conveying canals in addition to evaporation. Erodibility of these soils is a result of their coarse texture and should be taken into consideration in their management practices.

Sands are generally suited for permanent vegetation. Loamy sands in the dry warm areas are suited principally for grasslands. Sandy loams can be used for cultivated crops to a greater degree than the coarser-textured soils.

The fertility problems are the results of the low content of soil fine fractions and organic matter as well as the leachability of the nutrient elements by water.

Several other factors have considerable impact on the sandy soil use among which are:

- The texture of the surface soil.
- Depth, texture and structure of the subsurface soil.
- Amount and state of decomposition of the soil organic matter.
- Topographs.
- Seasonal and annual temperature rainfall and wind velocity.

Application of Clay or Organic Matter to
Decrease Loss of Water by Infiltration

Fine particles; the clay and silt fractions and organic manures are directly or indirectly applied to the soil. This practice is widely known and applied in sandy soil regions, especially in small holdings.

Although application of organic matter to the sandy soils has the same effect as application of the clay, usually organic matter is decomposed and has to be added frequently. Ahmed[4] showed that mixing 130 g/pot of Nile silt or 130 g/pot of wheat straw with the surface 15 cm of the sandy soil in a pot experiment increased the dry weight of maize plants from 88.7 to 150.6 g/pot and of barley plants from 38.5 to 62.3 g/pot, respectively.

These practices improve the physical properties of the sandy soils. They decrease the loss of applied water and nutrients. Their costs are too high to be considered as fertilization.

Organic matter is applied below the soil surface at different depths; this subsurface layer is not easily decomposed and remains, forming a relatively less permeable layer for about 10 years as reported by Egerszegi.[5] The layer decreases the water movement downward. Other materials were also suggested for forming subsurface layers at a depth of 50–100 cm. Egerszegi[5] placed one or more layers in the soil at least 1 cm thick, at depths from 38 to 75 cm, consisting of manure or compost.

Makled[6] conducted experiments on Southern Tahreer sandy soil in Egypt including the following treatments:

1. barriers of clay layers at 50, 60 and 70 cm deep;
2. farmyard manure layers at 50, 60 and 70 cm deep;
3. a perforated 3 mm thick asphalt layer at a depth of 60 cm + farmyard manure mixed with the soil just above that layer;
4. a 3 mm thick asphalt layer over a layer of cement bag paper at a depth of 60 cm + farmyard manure mixed with the soil just above that layer;
5. a perforated 3 mm thick asphalt layer and linen at a depth of 60 cm + farmyard manure mixed with the soil just above that layer;
6. a perforated nylon layer at a depth of 60 cm + farmyard manure mixed with the soil just above that layer.

He reported an increase in the yield from all treatments when compared with the control varying from 116 to 290%.

Melted bitumen was frequently used to form a layer underneath the soil surface either alone or sprayed over parchment paper to restrict the water movement in the soil profile.

Erikson et al.[7] suggested using layers of the bentonite clay or plastic sheets at variable depths of the soil. They pointed out that a continuous layer of these materials is difficult to establish. A layer 2–3 mm of asphalt 6000 liter/acre (14000/ha) can be formed using special machines. They showed from their experiments on sandy soils in Canada that the water retained by the soil at 20 cm above that barrier was about 30%. Without the barrier the water retained was only 10–12%.

Utilization of Soil Conditioners

In 1951 the term "Soil Conditioners" was introduced to mean the "Newly Born" chemicals that would improve the soil physical properties. At that time, the main problem of the fine clay soils and the nonsaline sodic soils, then known as "Black Alkali" soils, was the dispersion of their clay particles. Thus, they acquire agronomically unfavorable properties as impermeability to air and water.

The chemicals used in experiments conducted early in the 1950s were not persuasive. Their effect on the sodic soils was limited and their economics were prohibiting. Other practices were effective and much cheaper.

Though the soil conditioners introduced in the 1950s and 1960s did not satisfy the agronomists, the producers went on introducing one compound after another. At the same time, soil scientists in various institutes examined the chemicals produced. They did not lose hope for obtaining a conditioner that economically improves the soil physical properties. They considered that changing these properties, especially the soil structure, opens vast and promising horizons.

Since the manufacturing of the Kriliums by Monsanto in the 1950s, hundreds of products have been patented.

Though the physical properties of the sandy soils are quite different from the fine clay soils, both suffer from their unfavorable texture, and their poor structure.

The polymers that are capable of absorbing water several times—up to 500—their weight, offer a means to improve the water retention in the soil system in the vicinity of plant roots. Among these "super absorbents" are hydrolyzed starch, polyacrylonitrile graft copolymer (H-Span or Superslurper), vinyl alcohol-acrylic acid copolymers, and polyacrylamides. Mostajeran,[8] Miller,[9] and Hemyari and Nofziger[10] used various compounds of these jell-forming materials and showed that the soil water retention had increased, thus increasing the available water to the growing plants.

Utilization of these compounds was a turning point in the practices of sandy soils' agricultural use.

Factors Affecting the Conditioner Efficiency Factors
Related to the Conditioner

Polymers are synthesized to be nonionic, cationic, or anionic. When a polymer is applied to the soil it is usually adsorbed on the surface of the soils' single particles. This adsorption alters the particles relation with water and ions in the surrounding solution.

Attachment of one polymer molecule to several particles may also take place. Thus the formation of interparticle bonds enhances the flocculation of a dispersed system or stabilizes existing unstable or metastable arrangements of particles. Though this is the general outlook of the conditioner-soil interaction, a clear understanding of the phenomena occurring in soil conditioning is required.

Several requirements should be satisfied in the conditioner to achieve satisfactory results in a reasonable time.

- Adsorption of a polymer depends on its charge. Theng[11] reported that positively charged polymers are adsorbed largely through electrostatic interactions between cationic groups of the polymer and negatively charged sites of the clay surface. Little adsorption occurs with negatively charged polymers due to charge repulsion between the polymer and the clay surface.

- The compound has to possess good adhesive properties.
- Soluble polymers should become insoluble upon incorporation in the soil. This is possible by several mechanisms such as complexation, cross-linking, and adsorption to clay particles. Thus, the formed soil aggregates will be stable in water. If the polymer remains soluble in water, it dissolves in water upon its application to the soil and the adhesive bond is broken.
- The suitable life time of the compound is variable. It can be as short as two months in the case of stabilizing the soil surface against erosion until the planted crop reaches a satisfactory size. Some conditioners may remain active for 5 years.
- The conditioner should be cheap.
- According to Schamp[12] large molecules infiltrate deeper in the soil aggregates than smaller ones.
- The compounds should not be phytotoxic.
- Higher molecular weight of synthetic polymers enhances both adhesion and infiltration of the polymer solution.

Factors Related to the Method and Rate of the Conditioner Application

The method and rate of the polymer application to the soil affect the polymer efficiency as well as its economic feasibility.

The conditioner is applied to the soil as a dry powder, granules, a solution, or an emulsion. The powder is distributed as homogeneously as possible on the soil surface, then the soil is moistened and tilled.

Terry and Nelson[13] applied granulated PAM to the dry surface of field plots at the rate of 200 kg per ha. The soil was tilled to the depth of 10 cm, then an additional 200 kg/ha PAM was applied and the plots were again sprinkled with 5 cm of water. The soil was allowed to dry for 7 days and was then tilled for the third time.

El-Sherif[14] applied the dry polymer in pots. The required amounts of PAM 20% were 125 kg and 400 kg per ha for 125000 vegetable plants and 400 citrus trees, respectively.

In the study conducted by Wallace and Wallace,[15] the conditioner was dissolved in water and poured over 500 g of soil in a container. The volume of the solution was adjusted to match the water-holding capacity of the soil. They stated that with this procedure, the aggregates present in the soil would be stabilized if the concentration of the polymer in solution was sufficiently high.

Experiments conducted by Tayel and Antar[16] and El-Sherif[14] to determine the effects of bitumen emulsion polyamid polyvinyl alcohol, polyvinyl acetate, and lignosulphonate at several rates of application on aggregate stability increased progressively with increasing application rate of polyvinyl acetate polyacrylamide and polyvinyl alcohol. Soil water retention increased with the rate of application at $pF=0$ and decreased at $pF=2.53$, but the changes were negligible at $pF=4.18$. Available water for plant and water loss through evaporation was decreased in the treated soil.

The super absorbents are usually mixed with the upper few centimeters of the soil in the field. In pot experiments, the soil is ground to pass a 2 mm sieve, mixed with Super Sluper or other chemicals at the needed rates of 0.1 to 0.8% by weight.

Presence of a high concentration of soluble cations in the soil or in irrigation water might be a factor decreasing the adsorption of the polymer on the clay surface. Work is still needed to understand the effect of salts on the supergells in sandy soils.

The effectiveness of soil conditioners on plant growth is controlled by other growth factors. Wallace and Abuzamzam[17] showed that the least response to soil conditioners application was when the soil was deficient in N and P.

Table 27.   Efficiency of irrigation methods.

Method of Irrigation	Practical Efficiency
Flooding	0.70
Pipes with gates	0.65–0.75
Movable pipes	0.75
Pivot	0.85
Small sprinklers	0.90
Drip	0.95

Source: Land Master Plan of Egypt.[9]

Methods of Irrigation*

The main methods of irrigation are surface flood irrigation (such as check basen, furrow irrigation methods), sprinkler and drip irrigation methods (Table 27).

Because of the fast infiltration rate of water in the sandy soils, the flood irrigation methods are not suitable for these soils. Utilization of surface flood irrigation results in excessive loss of water and lower conveyance efficiency which may reach as low as 50%.

Surface irrigation method is not recommended for soils having an infiltration rate of 10 cm/h or more.

When surface irrigation is used, loss of water may be decreased and the irrigation efficiency is increased with the utilization of pipes. In the case of irrigating orchards the areas surrounding the trees only may be irrigated. The remaining area is left dry.

The open canals conveying water to the field should be lined. Loss of water from lined canals should not be more than 30 l/m²/day, while the loss from unlined canals crossing sandy soils may reach as high as 500 l/m²/day.

Lining the canals in sandy soils has several advantages:

a.   Protects the adjacent land from salinization.
b.   The area of the cross section of the lined canal is less than if unlined.
c.   Silting is also less in lined canals.
d.   Lining protects the canal itself from being eroded.

Several materials are suggested for lining canals:

- Concrete: It lasts as long as 45 years though it is more expensive.
- Cement-sandy soil mixture.
- Cement-sandy soil-plastic mixture.
- Polyethylene and butyl rubber.

Sprinkler Irrigation

This method of irrigation is recommended for sandy soils and soils having excessive gradient. It saves about 30% of the applied water as compared with flood irrigation methods. It also saves the cost of land leveling.

---

*   The methods of irrigation are discussed in Chapter 5.

Generally, sprinkler irrigation is the most suitable system for sandy soils. This system allows uniform penetration of water into the soil and reduces the danger of erosion by runoff water.

The main disadvantages of sprinkler irrigation methods are:

- They require high capital.
- Yearly depreciation may be high.
- Working cost is higher than surface flood irrigation method.
- The water distribution on the field surface may be disturbed with winds.
- Rainfall may form a surface crust which decreases water infiltration.
- They require high energy input.

Drip (Trickle) Irrigation

Plastic pipes are extended along the rows of plants. Water drops flow at the rate 2–10 l h^{-1} to form a moist soil circle around the plant. The distance between the pipe holes is adjusted to allow the contact of the outer circumferences of the moist circles. The distance between the tubes (the plastic pipes) is controlled according to the method of planting and the kind of crop. Though water has to be filtered before its use in the drip irrigation method, emitter clogging is a disadvantage.

It is worth mentioning that salts accumulate at the outer boundaries of the moistened circles. Leaching of the salts is recommended.

Drip irrigation saves water and manual work is very limited though capital cost is high. Yield per unit of irrigation water is generally higher than in case of other irrigation methods.

Drainage of Sandy Soils

Drainage of sandy soils is a controversial issue. Experts do not recommend drainage for these soils as they suffer from an excessive water infiltration rate and the main function of drainage is to get rid of excess water.

Other experts do not generalize this recommendation and prefer to examine the soil and give a suitable recommendation for each case.

Drainage may be recommended when:

- The soil profile contains hard pans. Rate of water infiltration is thus controlled by the permeability of the hard pan.
- When the examined field lies at a lower level than adjacent fields.
- The drains are necessary to receive rainfall especially when the field has a steep slope.
- When the land has a groundwater table, drainage is a necessity.
- Tile drains are the suitable system of drainage. Open drains are subject to collapse.

Controlling Soil Moisture Evaporation

Plants growing in the sandy soils are short of available water. Any practice that increases moisture in the root zone is recommended. Thus, decreasing the evaporation of soil moisture, though evaporation takes place in sandy as well as fine-textured soils, helps the plant to absorb its need of water.

a) Mulching is the widely practiced operation in the field to decrease soil moisture evaporation. The mulching material can be plant residues, paper, polyethylene, asphalt, or even by peples. According to Black and Grebs,[18] a complete or partial mulching of the land surface by plastic cover decreased soil moisture evaporation and improved crop utilization of water as compared with unmulched sandy soil surface. They stated that mulching with plastic accumulated nitrates in the soil.

Stubble mulching is a system of farming in which tillage, seeding, and harvesting are performed with the aim of keeping crop-residue cover on the surface of the soil.

The stubble should be left—after harvesting—standing (erect) as much as possible. In this manner it is as twice as effective in controlling wind erosion as flat stubble.

Maintaining a crop residue cover would (USSWC Res.[3]):

- protect the soil from erosion,
- reduce sealing the surface soil resulting from raindrop striking bare soil, and
- improve water penetration through the soil.

b) Eradication of weeds eliminates the loss of water by transpiration.

c) Plowing the soil surface decreases the water capillary movement to the soil surface.

d) Several chemicals are introduced to decrease moisture evaporation from the soil surface and transpiration from plants' leaves. Thus, utilization efficiency of water can be improved. Among these chemicals which decrease water evaporation from the soil are long chain alcohols such as Hexacandol and Deca sonol which can be mixed with the soil surface.

Chemicals that decrease transpiration from the plant such as undecanoic acid, phenyl-mercuric acetate, Vaporgard, and Epoxylinseed oil should be added in a suitable concentration, otherwise yield of crop is decreased. Also, some of these chemicals are poisonous.

Protection from Erosion*

The sandy soils are more subject to wind erosion than other soils. Sands and loamy sands are more susceptible to erosion than sandy loams. Fine and very fine sands, loamy sands, and sandy loams are more susceptible than medium and coarse. The texture of the subsurface soil has no immediate bearing on wind erodibility (USSWC Res[3]). Several causes aggravate soil erodibility:

- These soils may be open deserts or coastal areas without dense natural vegetative cover or objects that impede wind speed.
- In addition, the soil structure is weak due to lack of cementing materials. Few soil aggregates or clods can be built, and those are readily broken down by weather and tillage.

Also, for the same reasons, these soils are subject to water erosion in regions of sloping lands. Because the sandy loams have a slower water infiltration than coarser sandy soils, they are more susceptible to water erosion.

---

*   See more about soil erosion in Chapter 6.

Chepil and Burnett[19] stated the following to protect the sandy soils from erosion:

- Tillage practices should be minimum. Lister or chisel plows are more suitable than mould board discs.
- Plowing should be perpendicular to the wind direction and carried out with the soil somewhat moist to form cloggs not easily transported by wind.
- The growing crops should cover the soil surface, especially during the windy periods.
- Overgrazing of pasture should be avoided. Windbreaks reduce up to 60–80% of the wind velocity at 7 m above the soil surface. Orientation of the trees rows and their density have to be considered to obtain successful results from the windbreak. The trees of windbreak share the growing crops their nutrients. Also, they shade the crops. These effects can be detected 15–20 m far from the rows of the windbreaks.
- Deep-plowing is recommended by the U.S. Soil Conservation Service[3] under the following conditions:

    i.   Sandy soils having a layer that can be reached with the utilized plow and contain 20–30% clay. Areas that contain less than 20% clay and constitute less than 10% of the total area of the land are not recommended for deep plowing.
    ii. If the surface 20 cm only of the soil is sandy while the rest of the soil profile is made up of fine particles.
    iii. When the soil is turned over, about one fourth of the turned depth is made up of fine particles.
    iv. The subsoil layer which will be turned over should be fine coagulated particles, having good permeability and not containing more than 2% $CaCO_3$, otherwise the erodibility of the soil by wind is increased.

The U.S. Soil Conservation Service[3] recommends the following practices for nonirrigated conservation cropping system:

"High sustained production and improvement of nonirrigated cultivated land may be accomplished principally by suitable crops and crop rotations stubble mulching, stripcropping, fertilization, terracing and contour tillage wherever they are suited. The conservation cropping plan depends on flexibility and providing for substitutes whenever they are needed."

Fertilization of Sandy Soils

It has been stated above that the main constituent of the sandy soils is the sand fraction. Accordingly, the fertility level of these soils is controlled by their clay and organic matter content. Nutrients applied to raise their low fertility level remain in solution, subject to loss by leaching with irrigation or rain water. The portion which might be retained by the soil is controlled by the soil content of the fine fraction and organic matter. The leachability of nutrients is further aggravated with the frequent irrigation required for these soils.

Farmers of sandy soils are aware of the problems on the whole and they try to improve the soil fertility as well as its ability to retain applied nutrients. Their efforts depend mainly on fertilization by organic manures and application of clay. Several other means have been suggested.

Organic Fertilization

The loss of nutrients in organic manures by leaching is slow because of their insoluble nature. Hence, they remain in the soil within the reach of the plant root systems for a longer period than the inorganic fertilizers.

In a pot experiment using the sandy soil, Ahmed[4] showed that the dry weight of maize and barley increased by application of 130 g manure/pot (37 tons/ha). The former increased from 88.7 to 132.3 g/pot and the latter from 38.5 to 62.4 g/pot. He also showed that nitrogen absorbed by maize plants increased from 1.23 to 3.15 g/pot. In a field experiment Makled[6] showed that fertilization with farmyard manure on the soil surface increased the yield of alfalfa 16% above the control. In lysimeters and field studies, Mahmoud et al.[20] obtained results showing that addition of winter or summer green manure or compost to the sandy soil in equal quantities had increased the soil organic matter, total nitrogen, and soil microbial flora. They also showed that in addition to chemical fertilizers, application of clover or groundnut green manure or an equal compost had significantly increased sesame and maize yields above the control or treatment with chemical fertilizers alone.

Phosphorus

Uncultivated sandy soils are generally poor in P, having 300–500 mg kg^{-1} total P and 3–5 mg kg^{-1} P extracted by the $NaHCO_3$ method. In the alluvial soils these values are about 1000 and 12–15 mg k^{-1}, respectively (Balba[21]). Good soil management which leads to an increase in the fine fraction and the application of organic manures would tend to increase the phosphorus content and, consequently, the soil fertility.

The loss of fertilizer P by leaching is not a problem in most soils. Phosphorus is usually immobilized upon incorporation in the soil, but because of the low content of clay and organic fractions in the sandy soils, a limited movement of applied phosphatic fertilizers takes place by convection with irrigation water unless Ca salts are present. Early work by Balba[21] showed that phosphorus applied to a column 15 cm long of El Khanka sandy soil was distributed along the column and the filtrate contained 3.16 mg P, constituting 12.4% of the added amount. The filtrates of columns of a clay loam soil fertilized with P did not contain any significant amount of P above the content of filtrates of unfertilized columns. Actually, all added P was recovered in the first 2.5 cm of the fertilized columns.

Obviously, the immobilization of P at the soil column surface increases with the increase in the fine soil fraction. Thus one expects P to move with water when this fraction diminishes, as in the case of some sandy soils. In columns of pure sand the recovered P in the filtrates reached 98% of the added amount (Balba[21]).

Under field conditions, applied phosphatic fertilizers are better distributed in the root zone and the plants have a greater chance to absorb their need of this element. Experiments carried out on P fertilization in sandy soils showed crop response to increments of P.

Potassium

The situation of potassium in sandy soils is similar to that of phosphorus. These soils in Egypt contain about 5.0 meq of total K/100 g and about 0.25 meq/100 g soil of exchangeable plus soluble potassium. The alluvial soils of the arid regions contain about 10.15 and 1–2 meq/100 g soil of the two forms of K, respectively (Balba[22]).

The problem of losing applied potassium fertilizers by leaching with irrigation water is not serious in sandy soils. Added K is usually held in an exchangeable form on the surface of the fine soil fraction. Although the cation exchange capacity of the sandy soils is usually of the order of 5–10 meq/100 g, applied potassium constitutes a very small fraction of this capacity and can be retained by the soil. However, as stated above, in the case of soils very poor in the fine fraction, leaching of applied K may take place. Application of potassium to sandy soils is recommended, especially for vegetable crops and orchards. Irrigation water content of K may be an indirect supply of this element to crops grown on sandy soils.

Nitrogen

As stated above under long, warm summer and low precipitation, the soils contain low organic fraction. The uncultivated sandy soils contain as little as 0.008–0.015% organic matter; therefore, their total N content is also low (0.0015–0.002%). In addition to their low content of N, the leachability of N fertilizers in these soils constitutes a major nutritional problem.

Nitrogen fertilizers are mostly in the forms of NO_3^-, NH_4^+, or urea. The nitrates are not retained by the soil colloids. The ammonium is transformed to NO_3^-. Urea is hydrolyzed to $(NH_4)_2CO_3$ and subsequently nitrified.

Balba et al.[23] reported that NO_3-N is the main form which moves most readily with water. They showed that leached N from soil columns fertilized with NH_4 or urea was in the form of NO_3^-.

Broadbent and Tayler[24] showed that urea moved less rapidly than nitrate. The results of Balba et al.[23] showed that 14, 7.5, or 67% of applied N as $(NH_4)_2SO_4$, urea, or $Ca(NO_3)_2$, respectively, was accounted for in the filtrates. They reported that migration of soluble substances with water depends also on the water-holding capacity of the soil and amount and frequency of water application.

Slow-Release N Fertilizer

Considerable attention has been given to the development of nitrogen fertilizers with controlled availability, because they supply nitrogen continuously over an extended period, thus avoiding the need for repeated applications of conventional water-soluble fertilizers. They also promise to (1) minimize over-consumption of N, thus upset of nutrient balance, (2) to reduce N losses by leaching which constitutes a groundwater pollution hazard, (3) to decrease gaseous losses of N, and (4) to reduce the hazard of injury from overapplication, particularly to seedlings.

Several technologies have been practiced to achieve this purpose such as producing less soluble chemical compounds rich in N. Capsulating, granulating, pelleting, or coating the N inorganic fertilizers decrease their contact with water.

Fertilizers with Controlled Solubility

Armiger et al.[25] reported that the overall efficiency of properly formulated ureaform materials equals or exceeds that of conventional nitrogen fertilizers in respect to long season crops such as turf. They also demonstrated that a single application of ureaform may be made at higher N levels than would be feasible with more soluble N sources. Brown and Volk[26] reported that a greater proportion of ^{15}N-labeled ammonium nitrate-N was recovered in the plants than labeled ureaform-N. However, labeled N was found in the soil after one year when ureaform was the source.

Several other N-carriers are produced to serve as slowly available N sources to plants. Among these are coated urea 36.0% N, thiourea (NH_2CSNH_2) 36.0% N, hexamine ($C_6H_{12}N_4$) 40.6% N, oxamide ($H_2NCOCONH_2$) 31.8% N, glycoluril ($C_4H_6N_4O_2$) 39.4% N, and oxidized N-enriched coal 20.1% N.

Beaton et al.[27] showed that apparent recovery of N from various sources decreased in the following order: coated urea (75%), ammonium nitrate (74%), thiourea (69%), oxamidefine (65%), urea plus thiourea (63%), hexamine (59%), glycoluril (49%), urea formaldehyde (41%), and ammonium salt of oxodized nitrogen-enriched coal (39%). The only sources from which N was recovered in the fourth harvest were urea formaldehyde, thiourea, coated urea, glycoluril, and fine oxamide. With the fifth harvest only three sources, ureaform, thiourea, and coated urea, were still supplying nitrogen to a crop of orchard grass. This number was reduced to only two, ureaform and coated urea, in the sixth harvest. A small amount of nitrogen was recovered from thiourea in the seventh harvest.

Shehata et al.[28] planted pots of washed sand to maize with increments of N as ammonium sulphate (AS), urea (U), thiourea (Thu), hexamine (H), or oxamide (O) (see Chapter 2, Figure 6). After 2 months the components of N balance sheet for each N carrier were determined. They concluded that:

1. Straight-line equations were established between absorbed and supplied N. The slope ratio method showed that the efficiency of N carriers relative to (AS) was AS : AS3 (applied in 3 portions) U : Thu : O, equals to 1: 1 : 1.3 : 1.1 : 1.3.
2. The average percent utilization of AS=62, AS3=75, U=76, Thu=20, H=86 and O=77.
3. The leached N at the rate of application of 45 mg N per pot was from Thu > AS > AS3 > U > O > H.
4. The N remaining in the sand after cropping, in the case of applying 45 mg N per pot was O > Thu > AS3 > AS > H.
5. The N lost in gaseous forms (not accounted for N), in the case of applying 45 mg N per pot, was Thu > O > AS3 > U > AS > H.

Foliar Spray

To avoid the loss of nutrients added to sandy soils by leaching, they can be sprayed on plant leaves. Several investigators have shown that nutrients sprayed on the leaves have the same efficiency as those added to the soil in heavy soils. In sandy soils spraying fertilizers have given better results.

Witters et al.[29] reported that foliar feeding was most successful on crops when soil fertilization had not provided the necessary control of N uptake, or on crops that were regularly sprayed with micronutrients, insecticides, or fungicides. Urea sprays are used on many vegetable crops. The amount of N applied is often in the range of only 10 to 20 kg/ha per application. Foliar feeding of field crops like wheat and maize is also possible. Frank and Viets[30] reported that urea was most effective because of its rapid penetration into leaves; half of an application penetrated the leaf in 1 to 6 hours.

Oertli and Lunt[31] studied the factors influencing the rate of release from controlled-release fertilizers by encapsulating membranes. They found that the release rate was largely independent of the pH of the element or the soil pH. An increase in temperature from 10°C to 20°C almost doubled the initial release rate. The release could be regulated very efficiently through the coating thickness. There was an effect of ionic species; nitrate and ammonia were given off more rapidly than potassium or phosphate. Their experiments demonstrated that inorganic fertilizers, which in aqueous solutions would be rapidly soluble, are released slowly after coating. Gradual microbiological breakdown of the coating material can almost certainly be eliminated as a mechanism for the release of the nutrients. A working hypothesis of diffusion as a possible mechanism has been further supported. As a first step, water would tend to move outward. During a three-month growing period, an efficiency of recovery ranging from about 25 to 45% by maize was obtained from a single application incorporated in the sand of coated ammonium nitrate.

Ahmed and Whiteman[32] used two slow-release pelleted N fertilizers manufactured by Esso Engineering and Research Ltd., known as EAB 3032 (17.7% N, 21 days to release 79% of its N) and EAB 3033 (18.0% N, 63 days to release 75% of its N), compared them with $(NH_4)_2SO_4$ as a source of nitrogen for the rice variety Bluebelle in Trinidad. The slow release materials were buried 5 cm in the soil after transplanting, while the $(NH_4)_2SO_4$ was applied and worked into the soil just before transplanting. EAB 3033 resulted in a yield of 2920 kg/ha, EAB 3032 produced 2060 kg/ha, and $(NH_4)_2SO_4$ 1195 kg/ha. Plants treated with $(NH_4)_2SO_4$ had absorbed all their N when they had reached only 50% of their final dry weight. At the same stage, plants treated with slow-release materials had absorbed only half of their final N content and they continued to absorb N until a much later stage in growth.

Attoe et al.[33] studied the fertilizer release from packets and its effect on tree growth. They found that the length of time required to release the fertilizer was directly related to the size of the packet and inversely related to the number of pin holes. The more soluble fertilizer constituents were released faster than the less soluble ones.

Balba and Sheta[34] formed pellets of $(NH_4)_2SO_4$ (AS) gypsum, heated at 70°C for 24 h. They compared the leachability of these pellets with $(NH_4)_2SO_4$ and $(NH_4)_2SO_4$-gypsum mixtures in sand columns. The columns were watered 9 times and each time an amount of water equal to the maximum water-holding capacity was added. Each filtrate was collected separately. They showed that the pellets dissolved gradually, while almost all the N of the added NH_4 was collected in the fourth filtrate and the pellets yielded about 25% of their N. After 9 waterings, the sand column still retained about 29% of the added N as pellets.

In sand pots they compared the leachability and N uptake by maize of $(NH_4)_2SO_4$ (AS) added as follows:

i. "AS" in single application after thinning.
ii. "AS" divided into 3 portions: 1/4 after thinning, 1/2 10 days after the first portion, and the rest 10 days after the second application.
iii. "AS" + equal weight of gypsum.
iv. Pellets of AS + equal amount of gypsum formed as above.
v. The same as iv but ground and used as powder.
vi. Control.

They stated that differences in leachability of AS treatments greatly affected plant growth and N absorption. The maize plants utilized more N from AS-gypsum pellets and AS applied in 3 portions (treatment ii) than from other AS treatments. The superiority of treatment iv was mainly due to pelleting and not due to the presence of gypsum since treatments iii and v gave lower plant weight and absorption. Pelleting decreased the contact between the fertilizer and water. Thus N release slowed down and consequently was less leached, giving the plants a better chance to absorb their requirement of N.

Control of Nitrification

Controlling the nitrification of fertilizers results in a decrease of the nitrate nitrogen which constitutes the major part of the leached N compounds. Goring[35] studied the basic biological activity of 2-chloro-6-(trichloromethyl) pyridine in soil. The results showed that it is highly toxic to the organisms which convert ammonium to nitrate, and has a low order of toxicity to:

    i.   organisms or enzymes converting urea to ammonium,
    ii. organisms converting nitrite to nitrate,
    iii. general fungus and bacterial populations,
    iv. seedlings of many plants.

The minimum concentration in soil required to delay the conversion of ammonium to nitrite for at least 6 weeks ranged from as low as 0.05 mg kg^{-1} soil to as high as 20 mg kg^{-1} soil. Increasing concentrations delayed conversion for longer periods of time. Goring[35] also reported that similar controlled nitrification of $(NH_4)_2SO_4$, NH_4NO_3, $(NH_4)_2HPO_4$, and urea was obtained in broadcast and band applications, except when leaching conditions occurred immediately after fertilizer application.

Atta et al.[36] compared six chemicals as nitrification inhibitors. Their results showed that AM (2-amino-4-chloro-6-methyl-pyridine) and HQ (hydroquinon) were more effective for NH_4 oxidation inhibition than other compounds, namely Tamaron, Gramoxon, Benlate, and Carbon disulfide which were ineffective even for short periods.

Micronutrients

Sandy soils are poor in micronutrients. Iron, manganese and zinc deficiencies are noticeable on citrus trees grown on such soils. Foliar spraying with these elements and others is recommended.

Selection of Crops for Sandy Soils

Suitability of crops for sandy soils is based on their morphological and physiological characteristics related to efficient water use, drought tolerance, decreasing water loss by transpiration, and to economical aspects.

The morphological and physiological characteristics of desert perennials stated above (in Chapter 1) constitute the main characteristics of crop suitability for sandy soils.

Because of the leachability of applied mineral N fertilizers, leguminous crops enrich the soil with N fixed from the atmospheric N. The texture of the sandy soils allows the peanut flowers to penetrate through the soil surface to form their fruits.

Crops having adventitious roots such as small grains, corn, and sorghum make use of the soil moisture in the upper soil layers especially in nonirrigated regions. Deep rooted trees and crops benefit from infiltrated water to the deeper soil layers.

High-priced crops are usually preferred for the sandy soils because of the relatively higher cost of production in these soils. Thus, fruits and vegetables for exportation are main products of sandy soils.

Among the crops that proved successful on sandy soils are barley, wheat, peanuts, lupins, sorghum, and sesame. Also, several fruit trees which have a low water requirement such as vines, olives, and figs are successful. Citrus and mango are grown on these soils in spite of their higher water requirements.

Nonirrigated Cropping

Suitable crops and crop rotations with stubble mulching, stripcropping, terracing, and contour tillage are the main operations to be followed in order to obtain high sustained production from dry farming the sandy soils. The recommended rotation under these conditions is small grains with intertilled crops, grasses, and legumes. All crops should be fertilized as long as rainfall is satisfactory. Sandy loam soil in drier climates may be fallowed provided the fallow is in strips not exceeding 50 m wide and the surface is kept covered with crop residue.

The rain season controls the farming system. If rain falls in winter, small grains and legumes are the main crops. The land is left fallow and covered with crop residues in summer. Regions with a rainy summer and a dry winter may grow corn and sorghum. In some regions cotton is grown on the better soils—mostly sands loams—and sorghum on the poorer soils—mostly loamy sands. Cotton does not suit sandy erodible soils as it gives little residue cover.

Irrigated Cropping

The need for irrigation is imperative in semiarid and arid regions. The methods of irrigation are discussed above. Corn, beans, potatoes, cotton, sorghum, peanuts, and sesame are successful crops on irrigated sandy soils. Shallow-rooted crops such as hay, pasture, and small grains require more frequent irrigation than deeper-rooted crops.

Fruit trees are well-suited for the sandy soils since moisture in these soils is more evenly distributed to a greater depth than in fine-textured soils. Thus, trees can develop deep extensive root systems.

Plants physiologically prepared to grow under drought or saline conditions and at the same time containing useful materials for nutrition or industries such as Jojoba *(Simondisia chinensis)* and Cassava (tapioca) castor beans *(Ricinus communic* L.) are usually recommended for sandy soils.

REFERENCES

1. U.S. Soil Survey Staff, Keys To Soil Taxonomy, AID, USDA, SMSS Tech. Monograms No. 6 3rd Printing, Cornell University, pp. 95–130, 1987.
2. Soil Map of the World, FAO/UNESCO, Soil Map of the World, 1977.
3. U.S. Soil and Water Cons. Res. in the Great Plains States. Misc. Pub. No. 902, 1963.
4. Ahmed, K.A.R., Effect of application of Nile silt and organic matter on the productivity of Southern Tahreer Soil. Thesis in Land Reclamation and Improvement, College of Agriculture, University of Alexandria, Egypt, ARE, 1967.

5. Egerszegi, S., Plant physiological principles of efficient sand amelioration. Agrok. es Talaj., 13:209, 1964.

6. Makled, F.M., Effect of deep manuring of sandy soils on the yield of alfalfa in El-Tahreer Province of the U.A.R. Agrok. es Talaj., 16:179–184, 1967.

7. Erikson, A.E., C.M. Hansen, and A.J.M. Smucker, The influence of subsurface asphalt barriers on the water properties and the productivity of sandy soils. Trans. 9th ISSS Congress, 331, 1968.

8. Mostajeran, A., Study of the Potential for increasing plant available water in soil by the use of Super Sluper. M.Sc. Thesis, Iowa State University, Ames, Iowa, 1976. Cited in Miller,[8] 1979.

9. Miller, D.E., Effect of H-Span on water retained by soils after irrigation. Soil Sci. Soc. Am. J., 43:628, 1979.

10. Hemyari, P. and D.L. Nofziger, Super Sluper effects on crust strength, water retention and water infiltration of soils. Soil Sci. Soc. Am. J., 45:799, 1981.

11. Theng, B.K.G., Clay-Polimer interaction, 1982. Cited in Helalia and Letey,[36] 1988.

12. Schamp, N., Chemicals used as soil conditioners, a survey paper, Proc. of Third Int. Symp. on Soil Conditioning, Ghent, 13, 1975.

13. Terry, R.E. and G.A. Nelson, Effect of polyacrylamide and irrigation method on soil physical properties. Soil Sci., 14:317, 1986.

14. El-Sherif, A.F. Editor, A Research Project on Sandy Soil Reclamation, Final Report, Acad. Sci. Res. and Tech. & FAO, 1987.

15. Wallace, A. and G.A. Wallace, (a) Effects of soil conditioners on emergence and growth of tomato, cotton and lettuce seedlings. Soil Sci., 141:313, 1968.

16. Tayel, M.Y. and F. Antar, Effect of conditioner and rate of application on aggregate stability, water retention and evaporation. Egyptian J. Soil Sci., 18:19, 1978.

17. Wallace, A. and A.M. Abuzamzam, Interactions of soil conditioner with other limiting factors to achieve high crop yields. Soil Sci., 141:343, 1986.

18. Black, A.L. and B.W. Greb, Nitrate accumulation in soils covered with plastic mulch. Agron. J., 54:336, 1962.

19. Chepil, W.S. and E. Burnett, USDA Production Res. Report No. 64, 1963.

20. Mahmoud, S.A., S. Taha, A.H. El-Damaty, and M.S. Moubarek, Effect of green manuring on the fertility of sandy soils at the Tahreer Province, U.A.R. I- Lysimeter experiments of virgin soils. Jour. Soil Sci. U.A.R., 8:129, 1968.

21. Balba, A.M., Penetration and Solubility of Phosphorus in Egyptian Soils (unpublished), 1951.

22. Balba, A.M., Potassium forms and sufficiency to plants of some Egyptian soils. J. Inst. Tropical Agric. (Germany), 5:19, 1968.

23. Balba, A.M., M.G. Nasseem, and S. El Massry, A preliminary study of soil fertility of the Nasser Project along the N.W. Coast of U.A.R. I—Factors affecting the utilization of nitrogenous fertilizers by plants. Jour. Soil. Sci. U.A.R., 9:25, 1969.

24. Broadbent, H.G. and K.B. Tayler, Transformation and movement of urea in soil. Mulch Agron. J., 54:336, 1958.

25. Armiger, W.H., K.G. Clark, F.O. Lunstrom, and A.R. Blair, Ureaform greenhouse studies with perennial ryegrass. Agron. J., 43:123, 1951.

26. Brown, M.A. and C.M. Volk, Evaluation of ureaform fertilizer using 15 N-labeled materials in sandy soils. Soil Sci. Soc. Am. Proc., 30:218, 1968.

27. Beaton, J.D., W.A. Hubbard, and R.C. Speer, The coated urea, thiourea, ureaform, hexamine, oxamide, glycoluril and oxidized N-enriched coal as slowly available sources of nitrogen for orchard grass. Agron. J., 69:127, 1967.

28. Shehata, N.Y., A.M. Balba, and T.V. Hai, Incubation products and balance sheet of nitrogen carriers applied to maize. Alex. Sci. Exch., 5:221, 1984.

29. Witters, S.H., M.J. Bukovac, and H.B. Tukey, Advances in foliar feeding to plant nutrients. Symp. on Fertilizer Tech. and Usage. Proceedings, pp. 492, 1963.

30. Frank, G. and J.R. Viets, The plants need for and use of nitrogen. Soil Nitrogen. Published by Am. Soc. Agron., Madison, Wisconsin, pp. 503, 1965.
31. Oertil, J.J. and O.R. Lunt, Controlled release of fertilizer minerals by encapsulating membranes. I—Factors influencing the rate of release. II—Effect on recovery. Soil Sci. Soc. Amer. Proc., 26:584, 1962.
32. Ahmed, N. and P.T.S. Whiteman, Comparison of ammonium sulphate and slow-release nitrogen fertilizers for rice in Trinidad. Agron. J., 61:730, 1969.
33. Attoe, O.J., F.L. Rasson, W.C. Danke, and J.R. Boyle, Fertilizer release from packets and its effect on tree growth. Soil Sci. Soc. Amer. Proc., 34:137, 1970.
34. Balba, A.M. and Sheta, T.H., Nitrogen balance sheet of $(NH_4)_2SO_4$-gypsum pellets and mixtures and urea formaldehyde applied to corn in pots of sand. Plants and Soil, 399, 1973.
35. Goring, G., The control of nitrification of ammonium fertilizers and urea by 2-chloro-6-trichloromethyl pyridine. Soil Sci., 93:431, 1962.
36. Atta, S.Kh., M. Ragab, A.M. Zayed, and A.M. Balba, Effect of six nitrification inhibitors on the nitrification process and nitrifying bacterial counts in soils. Alex. Sci. Exch. Vol. 6, 4:399, 1985.
37. Land Master Plan of Egypt, Draft Final Report, Vol. 1. Main Report, Euroconsult-PACER, Ministry of Development, Cairo, 1985.

CHAPTER 5

Water Management in Arid Ecosystems

INTRODUCTION

Since his first step on the earth, man had realized that his life as well as his animal's and his plant's are tied with the presence of water. From this bare fact he started to secure water for himself and his crops and animals the year round. He searched for this precious material on the land surface and underground and even on tops of mountains. He taught himself how to store and to keep it clean. He tried to apply progressive technology in obtaining it without any contamination.

There are several regions on the earth which enjoy a high precipitation rate. The land becomes green and the crops satisfy the needs of the people. Water under these conditions is a common material. People hardly give water much thought unless it falls in certain seasons and does not fall in others. Accordingly, water becomes a main concern when people need to find drinking water in the dry seasons.

Still there are regions covering vast areas of the earth that receive unsatisfactory rains, if any, throughout the year. Under such conditions water becomes the essence of life. It controls man's life and his relationships with his neighbors and all his economical as well as civic activities. The efforts of man to find a source of water, store and efficiently utilize it, have become his main concern and activity in these arid regions.

TECHNOLOGIES FOR INCREASING WATER RESOURCES

Springs

Water flows from springs. Man had discovered that this water comes through cracks in the rocks from an underground source. He tried to keep the water clean by surrounding its site with a fence and covering it with a lid. The water is directed via lined canals to where it will be used for drinking, irrigation, or other purposes. Monitoring the water quality is usually carried out by analyzing samples taken periodically.

Collecting (Harvesting) Rain Water

Rain water harvesting have been practiced since ancient times by the Arabs (Al Anbat) in the Negev, the ancient Lebanese (Phoenicians), and the inhabitants of the southern coast of the Mediterranean Sea, from Egypt westward to the Atlantic Ocean (Shafeil[1]). According to Zaunderer and Hutchinson,[2] flood water farming has been practiced in the desert areas of Arizona and northwest of New Mexico in the United States for at least 1000 years.

In the Khadin system of India, flood water is imponded behind earth bunds and crops are planted into the moistened soil after infiltration of water. Collecting water from the roofs or directing the rain water to pits filled with sand and pebbles to decrease its evaporation and help in cleaning the water as a filter from any floating material were practiced by man in various communities (Garduno[3]).

In other communities, the pits are lined with butyl or synthetic rubber to collect the rain water. The pits can also be lined with cement or with stone. The soil surrounding the pit may be treated with chemicals to become impermeable for water so that water moves faster toward the storage. Also, sheets of polyethylene or rubber can be used as a cover for the area surrounding the pit (Garduno[3]).

Water points in the deserts are stores of water that are used by the livestock during grazing or transportation to the market.

Rain water storing has several techniques. In some communities rain water is directed to a plain (above ground) surrounded by hills made by scraping the land around the area. Crops are planted after infiltration of the impounded water. Reservoirs of river water are built to store it to be used when water is needed. Some of these reservoirs are made to control floods and store water such as the High Dam of Egypt which controls the Nile flood (Shafeil[1]).

Fog Harvesting

Some inhabitants in South America use a fog catcher, 2.5 m high connected to a series of nylon strings 1 mm in diameter and 1.2 m long. Fog is condensed on these strings and collected. It is said that this gadget collects 18 L h[-1] (Garduno[3]).

Underground Water

The underground water is about 22% of the total amount of fresh water of the earth. This water concentrates in specified layers capable of storing it. It is possible to lift it by digging wells of various depths or, sometimes, it flows under piezometric pressure.

Man has made several kinds of wells (Garduno[3]).

Vertical Wells

The ground is dug vertically until the layer carrying the water is reached. The water is present under pressure and usually flows vigorously to the ground surface, or it may be obtained by using a bucket with a rope or an electric pump.

Wells may also be dug in the bottom of a river. Probabilities of water infiltrating underneath the river-bed are high. In deserts wells are the sole source of water for man and his animals. Man has used wells since ancient times. Our well digging tools have evolved to the point where we can now dig wells several hundred meters deep that use electric pumps and are equipped with networks of pipes to distribute water.

Horizontal Wells

Digging might be done horizontally in the mountains to reach the sites where water is stored. These wells mostly do not require lifting.

Quanats

Aflag—plural of falg—are canals connecting the well and the site where the water will be used. They are mostly underground. Thus, the falg is a tunnel 2–3 km long and may reach 100 km long. These aflag are well-known in the southeastern part of the Arab Peninsula, Iran, and Iraq. In Oman there are some eleven thousand of these "quanats," 7000 of which are in use. The discharge of these quanats may reach as much as 20–60 m^3 per second to irrigate 20–60 ha (Garduno[3]).

Wind Wheels for Lifting Underground Water

Wind power is used in many locations to lift water from wells. The wind wheels must be designed to start and work in low wind speeds and be equipped with a governing device to protect them during high winds. They must also have storage tanks to contain the pumped water. Wind power is cheap but limited in output and variable due to shifts in the weather (Garduno[3]).

Transporting Icebergs

This is a recent technology which is not practiced by man yet. The floating icebergs are made up of fresh water that is almost distilled water. The crumbs of the Antarctic lose yearly more than 10×1012 m^3 of ice in the form of floating icebergs which melt and disappear. One of the largest known icebergs was 250 km long, 90 km wide, and 350 m thick. Its area was as much as the area of Belgium. Such an iceberg was accumulated and pressurized in the Antarctic during several thousands of years (Victor[5]).

An iceberg 1200–1500 m long and 300–400 m wide which weighs about 100 million tons may lose about 20% of its size during its transportation.

The Antarctic enjoys more attention concerning transporting floating icebergs. Victor[5] stated that this attention is due to:

- Most of the floating icebergs of the north pole are not stable and have irregular shapes.
- The icebergs of the north pole are made up of several crumbs and do not form icebergs of suitable sizes, while those of the Antarctic are large and regularly shaped.
- The ocean currents from the north to the south do not suit the transportation operation as do those of the south pole.
- The iceberg slides very slowly from the edge of the continent or it may form a shaped icegulf 250–350 m thick.

Feasibility of Iceberg Transportation

The International Conference for discussing the difficulties encountered in iceberg transportation (Iowa, USA, 1977) proved that such an operation is feasible by utilizing modern technologies at a cost 30–50% lower than desalting sea water. Expenses until Giddah (Saudi Arabia) were 60 cents/m^3 while desalting expenses were 80 cents/m^3.

Among the difficulties encountered in the transportation operation are protecting the iceberg from erosion, and melting and evaporation as a result of friction with the sea water.

It was suggested that covering the iceberg with a reflecting material would decrease melting and evaporation of the floating part. To protect the part under sea water, it was suggested to cover it with a curtain equipped with a water basin.

When the iceberg reaches the harbor, it remains about 10 km away from the shore. Water is obtained from the melting iceberg to the shore in a pipe line leading to a storage unit on the coast. A network of pipes distributes the water.

Rain-Making

Although man has known much about rains and clouds, he could not, for a very long time, know how to increase the amount of rainfall in a specified location or to have the rainfall at a specified time.

Man knew that the water vapor does not fall as rain unless it is condensed to droplets which usually takes place around a crystal. The dream was to force the water vapor—the clouds—to form droplets by supplying the cloud with the seeds—the crystals—around which the droplets can be formed.

It was in 1946 when Schaefer and Langmuir of the General Electric Co. made the first trial. They used silver iodide whose crystalline structure is similar to that of natural ice. Thus it can be used as a nucleus for the droplets. The other material was dry ice, the frozen carbon dioxide.

The crystals of silver iodide are sown by an airplane or by a gun on land toward the cloud. The dry ice is usually sown from an airplane.

Rain-making attempts were received with great enthusiasm by the farmers who thought that it could save their crops from drought. The suggested methods were not always successful. An evaluation of these methods showed that rainfall in sites where attempts were made to obtain rain did not exceed more rainfall than in other sites where no attempts were made. However, several examples of success in rain-making were cited by UNESCO Courier[4] in Florida, USA. Also, in the UN Org. for Meteorology report the USSR was able to increase the rainfall in Ukrania by 13–18%. Other examples were in Australia and the Philippines. The UNESCO Courier concluded that obtaining more rain through injecting the clouds did not reach the success realized by injecting the fog with silver iodide or dry ice to condense it in airdromes.

Desalting Saline Waters

Evaporating the water and condensing the vapor to liquid water has been practiced for a long time, especially in laboratories, on ships, and in some harbors which do not have a satisfactory source of fresh water.

Other techniques to separate the dissolved salts from the water are also practiced such as electrodialysis using semipermeable membranes. Also, by freezing, ice is crystallized separately from salts.

Because any of the methods requires energy, several sources of energy were investigated. Though nuclear energy would be cheaper, the capital investment is an obstacle. Desalting of brackish surface and ground waters by osmosis, ion exchange and electro-dialysis, are both economically feasible for drinking water but not for agriculture (Garduno[3]).

MANAGEMENT OF RAIN-FED AGRICULTURE

The major portion of the world agricultural production depends on rainfall.

In humid and subhumid regions, water is not limiting and does control agricultural activities. On the contrary, in arid and semiarid regions agriculture is limited to areas and seasons that can provide sufficient water for the grown crops. Man has developed technologies which allow him to obtain reasonable yields of crops; otherwise, these areas would be left for pasture.

Planning for rain-fed farming requires:

1. A survey for the land and water resources.
2. Evaluation of the sufficiency of water for the need of the selected crop or crops.
3. The selection of crops is based mainly on the sufficiency of the evaluated sources of water, the period of precipitation, and the climatic conditions in the rainy season. Thus, if the rainy season is winter, wheat, barley, faba beans, and other winter crops will be grown. If summer is the rainy season, maize, sorghum, cotton, and other summer crops will be grown.
4. A plan for the culture operations which suit the land, climatic conditions, and selected crops will be set.
5. The success in rain-fed agriculture depends on sound management that efficiently utilizes all available sources.

A basic principle is to depend on surveying, evaluating, and monitoring the land and water sources by well-trained specialists.

In rain-fed agriculture, sustainability of crop production wholly depends on the amount and distribution of rainfall during the growth season of the selected crop and its water requirement. In Syria the production sustainability is divided into the following levels (AOA[6]).

First Sustainability Region

The average rainfall is more than 350 mm in the growth season and is not less than 300 mm in 66.6% of the recorded years. This means that production is guaranteed in 2 out of 3 seasons. This region is considered the lowest limit for acceptable sustainable wheat production in rain-fed agriculture.

Second Sustainability Region

The average rainfall is 250–350 mm and is not less than 250 mm in 66.6% of the recorded years. Thus a yield of barley is guaranteed in 2 out of 3 years. This region is considered a second degree of sustainability for wheat production but the first for barley.

Third Sustainability Region

The average rainfall is 250 mm and is not less than 250 mm in 33.3–66.6% of the recorded years. Sustainability of this region is low. Barley may produce a yield once or twice every 3 years.

Fourth Sustainability Region

Rainfall average is 200–250 mm and is not less than 200 mm in 50% of the recorded years. This region is suitable only for barley in fine-textured soils and usually with low yield or may be left for pasture.

Fifth Sustainability Region

This is not suitable for rain-fed cropping. It is left for pasture.

Rotation in Rain-Fed Regions in Syria

In regions known for their high production of wheat, the wheat is followed by fallow during the summer after harvesting. The fallow is extended to the following winter and summer seasons. The land is prepared for wheat in autumn. This signifies that wheat is planted once in 2 years. In regions where rainfall average is less than 250 mm, the area left fallow is increased to 3/4 or 4/5 of the land which signifies that only 1/4 or 1/5 of the area is planted.

Leaving the land fallow is based on the belief that the soil recovers its fertility and conserves moisture from the precipitation of the following winter season.

A study of this belief conducted by Loizidis,[7] an FAO expert in Syria in 1968. He compared the following rotations:

1. Wheat after wheat yearly.
2. Wheat followed by fallow.
3. Wheat followed by lentis.
4. Wheat followed by vicia (a leguminous forage crop).

He determined the soil moisture to the depth of 120 cm in autumn (November), the beginning of the spring (March), and before harvesting (May). The study showed the following:

1. The differences in soil moisture in November, March, and May for each rotation explain the differences in wheat yields.
2. The yield of wheat after fallow was better than after wheat.
3. Wheat after lentis is comparable to wheat after fallow.
4. Wheat after vicia is almost equal or better in some observations than wheat after fallow.
5. The soil fertility is improved after fallow.

The same conclusions were arrived at in 1978 by the Syrian Ministry of Agriculture.[8]

Culture Practices in Rain-Fed Farming

1. After harvesting the wheat, the soil is worked by scratching with a 6-blade scratcher to get rid of weeds and to form a friable soil surface layer to decrease soil moisture loss through vaporization. This machine leaves wheat straw in the soil to protect the soil from wind erosion.

2. In autumn the same operation is repeated after equipping the machine with blades 6.5 cm wide which reach 30 cm deep in the soil. This operation opens the soil so that rain water can easily penetrate it. The wheat straw is left upright to protect the soil from erosion.
3. The soil is worked in spring with the same machine for the same purposes for weed control.
4. The wheat seeds are sown in rows using the sowing machine following the plowing blades.
5. Fertilization is practiced but it needs improvement and further studies.
6. In regions receiving less than 300 mm of rainfall, selection of the catchment areas which receive runoff water from the surrounding elevated areas supplies the wheat in the low-lying areas with sufficient water.

In the North Western Coast of Egypt, average annual rainfall is 150 mm (Balba and El Gabaly[9]). Success of rain-fed agriculture depends on the selection of the area to be sown by barley or, more recently, by wheat. Areas receiving runoff water have better chances to produce yields, though much lower than irrigated wheat or barley.

Low-lying areas surrounded by elevated heights are usually planted with olive or fig trees. These areas have deep soil profiles as they receive fine-textured deposits with the runoff water. Rainfall water is directed to stores cut in the land and lined with cement to provide water in summer in the early growth stages of the trees. Olives and figs' water consumption is known to be low.

Areas which do not receive runoff water or which have shallow profiles that do not suit field crops or fruit trees are usually left for pasture.

The Shifting Agriculture

In dry regions where few crops can survive, the land is cleared of aboveground vegetation only. Much of the mineral nutrients (especially phosphorus) bound in this material is recovered by burning it or allowing it to rot in place, then followed by mulching the soil surface with vegetation and only lightly tilling it with hoes. After two or three, and sometimes 10 seasons, the yields become low and weeds increase. The farmers usually abandon the plot for a new one. The old plot is left to recover its fertility in 5–10 years. During this period, trees and shrubs grow again and bring up nutrients from the depth. The recovering vegetation adds litter to the soil to build up its organic matter and nitrogen content. Grass roots bind the soil, helping it to recover. Its erosion is controlled (Kampen and Burford[10]).

Because of the rapidly increasing costs of developing irrigation projects and the fact that many regions of the world do not have water resources for irrigation development, awareness of the importance of dryland farming has gained importance among soil scientists and farmers. Recognizing the role of high-yielding varieties of wheat, rice, and corn in increasing food production in several countries, dryland farming requires cultivars characterized by drought resistance and/or salt tolerance. Also, plants that can thrive on phosphorus, or nitrogen, deficient soils are needed.

IRRIGATION

In arid ecosystems, irrigation is a necessity for sustainable agriculture. For dependable irrigation, several requirements should be satisfied:

i. A source of water supplying sufficient amounts of water at the suitable time for the plants.

ii. A system for conveying the water from its source to the field and another system to distribute the water among the plants in every part of the field.

iii. An efficient system for drainage.

The previous pages have pointed out technologies to acquire sources of water that satisfy known needs. Water points supply drinking water for man and his herds in the desert, while large reservoirs supply vast areas with water for agricultural use the year round.

Variable investigations must be conducted for irrigation planning by surveying the area taking into consideration the soil characteristics that may affect the efficiency of the irrigation and drainage processes such as (U.S. Bureau of Recl.[12]):

1. Characteristics of the soil surface which require special study includes:

    * general land form,
    * soil texture,
    * land slope,
    * microtopography,
    * flooding,
    * evidence of erosion or deposition, and
    * surface rockiness or stoniness.

2. The surface infiltration rate should be determined for each soil mapping unit. Thus, hydraulic conductivity needs to be measured quantitatively in the field using water of the same irrigation water quality.

3. Amount, kind, and distribution of clay minerals are important to water movement, retention, and availability to plants.

4. Other soil water constants need to be determined in the field using the same water to be used for irrigation.

5. Planning an efficient drainage system requires extending the soil investigation to a minimum depth of 3 m. Impermeable, slowly permeable, and salinity of soil layers must be identified.

6. If paddy rice is to be grown on the soil to be irrigated, the soil capacity to maintain water on the soil surface and the infiltration rate are important. If dry crops are to be grown in rotation with the rice, the likely fluctuations of the water table, the water-holding capacity, and the drying period of the soil need to be determined.

The *quantity* of water available for irrigation should match the need of the selected crop in the area to be irrigated. This can easily be calculated. The *quality* of the water should also be known so that selection of crops and amount of applied water can be decided on sound basis.

Several means were practiced to convey water to the field. Among these are:

* Canals lined or unlined of different capacities.
* Pipes having various lengths and diameters.
* Pumps to lift the water or pressurize it in the canals, pipes, sprinklers, or drippers.

The Selection of the Irrigation Method

Surface flooding or sprinkler or drip irrigation methods should suit the soil texture, topography, slope, and the depth of its profile. Sandy soils having hydraulic conductivity of 10 cm/h should not be irrigated by surface flooding. Also, there are plants that do not tolerate direct contact with water such as citrus trees or watermelon plants. Culture practices as well as kinds of machinery also need to be considered.

The following points should be given considerable attention:

1. The soil profile should be well examined. Presence of an impermeable layer, or even with lower permeability than the layers above, leads to slowing water movement through the above layers and to water accumulation above the impermeable layer causing waterlogging.

   According to the U.S. Bureau of Reclamation,[12] the slowly permeable stratum is considered a drainage barrier when it has a permeability value less than one fifth the average of the horizons above it.
2. Parts of the area having different elevation levels should be separated by drains to protect the lower part from water seeping from the elevated part.
3. Canals should be lined in coarse-textured soils.
4. Leveling is necessary if surface flood irrigation is used.
5. The areas of basins in the sandy soils should be small to lower water losses. Canal openings should be narrowed to permit better water penetration through the soil.
6. In furrow irrigation, it is recommended to pre-irrigate the area to decrease salts.
7. Frequency of irrigation is increased in salt-affected soils.
8. The root system of plants in their first growth stage is small. Thus, irrigation needs are usually small at this stage. Later, plant water consumption increases with growth until maturity. Irrigation should be adjusted to match the plants' need.
9. Irrigation with water containing salts requires applying an additional amount of water—the leaching requirement—to avoid salt accumulation in the soil root zones, Richards.[13]
10. The time of irrigation is an important point. There are several techniques to decide when to irrigate the land.

When to Irrigate

- Before sowing, a heavy irrigation is made so that the soil is moistened to a suitable depth depending on the soil texture and kind of crop. In other methods the seeds may be sown on dry land, followed by irrigation.
- Irrigation is repeated when the soil moisture reaches about 50–60% of the available soil water in the soil layer containing the majority of the root system. The available soil water has customarily been regarded as equal to the soil moisture content at the "field capacity" minus the soil moisture content at the "permanent wilting point."

This concept signifies that this range of water (FC-PWP) is equally available to plants (Viehmeyer[14]). Thus, irrigation may be delayed until most of the available water has been consumed and the soil moisture has dropped very close to PWP which implies infrequent irrigation with larger amounts of water. However, this concept is not generally accepted. Another concept postulated earlier by Wadleigh (1955)[15] states that plant growth is a function of soil moisture stress. Thus, irrigation is scheduled whenever the moisture tension approaches the level at which plant growth is affected (Hagan[16]).

Gardner[17] concluded his available water concept by stating that a single pair of water potential values is not adequate to characterize the upper and lower limits of water availability for all soils and all crops.

The depth and textural variations of the solum horizons affect the moisture holding-capacity in the field. Water tension throughout this depth in the field may be measured using a tensiometer. Also, water content may be determined on soil samples taken from several depths or by using a neutron probe.

The nature of crops to be grown determines the depth of soil to be examined for assessing the available soil water. Examining a depth of 120 cm is recommended.

The total available moisture is the summation of the contribution of each horizon within the total depth considered (FAO Soil Bull.[18]).

The following examples (FAO Soils Bull.[18]) illustrate the way in which data on available water can be set out and developed:

a)   For a single horizon:

    texture   : loamy sand
    depth (0–30 cm) : 30 cm
    field capacity   : 8.8% by weight
    PWP (15 bar)   : 3.2% by weight
    bulk density   : 1.55 g/cm^3
    available water : (8.8–3.2) × 30 × 1.55 = 2.6 cm

b) For a soil comprising four horizons:

Depth	Texture	Available Water, cm
0–30 cm	Loamy sand	2.6
30–55 cm	Loamy sand	2.2
55–90 cm	Loamy fine sand	3.1
90–120 cm	Medium sand	1.1
		9.0

In a uniformly wet soil profile, plants initially extract water from the upper horizon, progressing downward through the profile. Rainfall or irrigation increases the moisture near the surface. Thus, the extraction zone goes back to the surface and moves downward again (Gardner[17]). Gardner[17] further showed that the shapes of the extraction from the surface downward represented by extraction curves after 10, 22, and 50 days for several crops were similar. He concluded that 40% of the extraction is from the upper 20% of the root depth, then 33%, 20%, and 7% from successive depths.

•   Experienced farmers depend on the plant appearance to decide when to irrigate. Some orchard growers depend on indicative plants such as sunflower to irrigate their orchards when they notice that these sunflower plants started to wilt.
•   Other farmers have the ability to decide when to irrigate from touching the soil, together with its appearance. They dig a pit in the soil layer that contains the majority of the root system and examine the soil with their fingers. The test is repeated in several pits and for several weeks. By experience they can decide whether the soil needs irrigation or not.

- Determination of the soil moisture at the depth of the majority of the root system is a direct method to decide the need for irrigation.

There are several methods for determining the soil moisture:

a. Tensiometers laid in the appropriate depth measure the moisture tension when equilibrium between the soil moisture content and of the tensiometer cup is attained.

b. Gypsum blocks determine the soil moisture content depending on measuring the electrical conductivity between two electrodes buried in the soil by a conductimeter. The electrical conductivity increases with the increase of the soil moisture content. Evidently this method does not suit saline soils or when saline water is used for irrigation. Plastic blocks are also used. They have the advantage of measuring the soil moisture content until the saturation percentage.

c. There are several other methods for determining the soil moisture depending on thermal soil properties or radioactive techniques, Jery et al.[19]

Recently, the time of irrigation and amount of water to be applied can be known by electronic calculators fed information about climatic records, soil properties, kind of plant, the last irrigation, and other culture practices (Hiler and Howell[20])

Methods of Irrigation

Surface Flood Irrigation

A network of canals to convey water from the source to the fields should be dug. Lining is a necessity if the canals cross coarse-textured soil. Pipes may also be used. The canals differ in their length, cross section, and slope according to their degree. Each land parcel should have a canal to supply water and a drain to receive surface and subsurface drainage water. The slope of the land should be capable of being graded to 1% or less (FAO[18]).

Several methods are used in surface flooding.

Furrow Irrigation

Water enters between furrows and extends horizontally to wet the furrow. The width of the furrow depends on the land slope, properties, and amounts of water as well as the shape of the field. This method is recommended when:

i. the amount of water is limited,
ii. a crust is formed on the soil surface after irrigation, and
iii. the land surface is not level.

The advantages are (Garduno[3]):

- Uniform water application.
- High water application efficiency.
- Good control of irrigation water.
- Control equipment available at low cost.

Wide furrows or terraces may be used to decrease plant contact with water. When the terraces are along the land contours, special care should be given to control water flooding from the edge of the terraces.

Basin Irrigation

The basins should be level. The basin may be divided by furrows for water control and its infiltration through the soil. Spacing of the furrows depends on the soil properties and its slope.

The advantages are (Garduno[3]):

- Good control of irrigation water.
- High water application efficiency.
- Uniform water application and leaching.
- Low maintenance costs.

Small basins may be required in the case of vegetables.

Surge Flow Irrigation*

Surge flow irrigation is an intermittent way of discharging irrigation water into irrigation pathways. The "on and off" periods may be constant or variable provided the sum of water applied satisfies the soil moisture deficit. According to Stringhamet[21] this method was introduced in 1979 by Stringhamet and Reller[22] in Utah State University as an attempt to automate the post-advance phase of an irrigation event. Studies showed that the permeability of the soil surface layer may be decreased. Intermittent flow had the potential for significantly improving the performance and versatility of surface irrigation systems. The increase in the efficiency of surface irrigation systems using intermittent application may reach that of the sprinkler or trickle irrigation. In addition, it made automation of surface irrigation possible and did not require water pressurization (Stringhamet[21]).

Sprinkler Irrigation

The main advantage of this method is that it allows irrigation of unlevel lands which require considerable expenses for leveling for flood irrigation methods (land slope 1–35%, FAO[18]). Because sprinkling saves water, it is recommended when water is limited. The rate of water application should not exceed the infiltration rate of the soil. Sprinkler irrigation method has several advantages such as (Garduno[3]):

- It suits the sandy soils having excessive hydraulic conductivity.
- It allows irrigating sloping soils without erosion.
- Water regular distribution on all parts of the field.
- Because leveling is not necessary, expenses of land preparation are low.
- It is possible to control the applied water to match the soil hydraulic conductivity.

---

*   Categories and description of "surge systems" are presented in details in Stringhamet.[21]

Sprinkler irrigation, however, has several disadvantages (Garduno[3]):

- The capital investments are high and at the same time the depreciation rate may also be high.
- Working expenditure is higher than the corresponding expenses of flooding methods.
- Wind may disturb the water distribution on the parts of the field.
- Crusts formed after rainfall may decrease water penetration in the soil.

There are numerous kinds and makes of sprinkler irrigation, varying from micro-sprinklers to pivot systems, fixed and movable systems. The movable systems are more common. They consist of pipes having nozzles. The pipes are made of galvanized steel 3–15 m long and 5–15 cm in diameter, according to the amount of water to be sprayed. The movable system depends on quick fitting between the pipes without leakage. Thus, the pipes can be moved from one place to another while the pump may remain in its site, or it can also be moved with the pipes. In semimovable systems, the pipes are fixed while the pump may be moved. In fixed systems, both the pump and the pipes are fixed in the field.

Toxicity Effects due to Sprinkler Irrigation

Sodium and chloride in the sprinkled water are absorbed by the plant leaves and may cause toxicity to the plant depending on the following conditions (Tables 28 and 29):

- High temperature, low humidity, and windy conditions increase absorption and toxicity.
- Rotating sprinkler heads present greatest risk. The water evaporates between rotations causing evaporation of water, thus salts are more concentrated in the remaining drops of water.
- Slowly rotating sprinklers (less than 1 revolution per minute) cause alternate wetting and drying cycles. The slower the speed or rotation the greater the absorption.
- Sprinkler irrigation more frequently.
- Daily sprinkling may cause problems.
- Toxicity to sensitive crops occurs at relatively low sodium or chloride concentration (>3 meq/l). It is expected that these crops are more sensitive to foliar Na absorption, Bernstein and Francois.[24]

The above stated factors affecting sodium or chloride toxicity during sprinkler irrigation led to suggesting practices that may reduce the adverse effect of sprinkler water containing Na or Cl.

1. Night sprinkling decreases or eliminates both Na and Cl toxicity due to foliar absorption. This is a result of increased humidity lowering temperature and decreased wind speed, Table 30, Busch and Turner.[25]
2. Hot dry winds aggravate the absorption and deposition of sprinkled water Na and Cl. Irrigation at night is a way to avoid these conditions, Busch and Turner.[25]
3. In hot windy areas, if the downwind drift from sprinkler irrigation lands on adjacent plant leaves, it is more concentrated than the applied sprinkler water. If movable sprinklers are used, they should be moved progressively downwind rather than upwind. In this case the sprinkler water washes away salts as soon as possible. Sprinkling in the early morning or late evening and night hours when the

Table 28.   Effect of irrigation method and water quality on the yield of tomatoes, t/ha.

Irrigation Method	EC of Water, dS m^{-1}	
	0.4	3.0
Drip	66.7	65.0
Sprinkler	52.0	39.2

Source: Abrol et al.[31]

Table 29.   Relative tolerance of selected crops to foliar injury from saline water applied by sprinklers.[1]

Na$^+$ or Cl$^-$ Concentrations Causing Foliar Injury[2] me/l			
<5	5–10	10–20	>20
Almond *(Prunus dulois)*	Grape *(Vitis* spp.)	Alfalfa *(Medicago sativa)*	Cauliflower *(Brassica oleracea botrytis)*
Apricot *(Prunus armeniaca)*	Pepper *(Capsicum annuum)*	Barley *(Hordeum vulgare)*	Cotton *(Gossypium hirsutum)*
Citrus *(Citrus* sp.)	Potato *(Solanum tuberosum)*	Corn (maize) *(Zea maya)*	Sugarbeet *(Beta vulgaris)*
Plum *(Prunus domestica)*	Tomato *(Lycopersicon lycopersicum)*	Cucumber *(Cucumis sativus)*	Sunflower *(Helianthus annus)*
		Safflower *(Carthamus tinctorius)*	
		Sesame *(Sesamum indicum)*	
		Sorghum *(Sorghum bicolor)*	

[1]   Susceptibility based on direct accumulation of salts through the leaves.
[2]   Leaf absorption and foliar injury are influenced by cultural and environmental conditions such as drying winds, low humidity, speed of rotation of sprinklers, and the timing and frequency of irrigations. Data presented are only general guidelines for late spring and summer daytime sprinkling.
Source: Data taken from Maas,[58] cited in Ayers and Westcot.[23]

wind speed slows down is recommended. Grouping sprinklers in blocks is preferable to long widely spaced single rows if drift is likely to be a problem, Ayers and Westcot,[24] Busch and Turner.[25]

4.   More frequent or continuous wetting of foliage allows less drying of leaves and less absorption than intermittent wetting and drying. A sprinkler head rotation of one revolution per minute or less is often recommended if it is possible, Ayers and Westcot.[23]

Table 30.   Sodium content in cotton leaves in percent oven dry weight.[1]

Variety	Day Sprinkled	Night Sprinkled	Surface Irrigation
Short staple	0.73	0.46	0.44
Long staple	0.29	0.12	0.10

[1]   Irrigation water quality EC_w = 4.4 dS/m; Na = 24 me/l.
Source: Data taken from Busch and Turner[25] (1967).

Table 31.   Leaf burn on alfalfa with three rates of water application by sprinkler irrigation in Imperial Valley, California.[1]

	Rate of Application (mm/hr)		
	1.8	2.7	4.0
Alfalfa plants with leaf burn (percent)	92.5	5.0	2.5

[1]   Irrigation water quality EC_w = 1.35 dS/m; TDS = 875 mg/l; Na = 6 me/l; Cl = 7 me/l.
Source: Data taken from Robinson[26] (1980).

5. The rate of water application may be accomplished by a) enlarging the sprinkler orifices, b) increasing the pressure, or c) reducing the spacing on the sprinkler system, Table 31, Robinson.[26]

6. Decreasing the moistened leaves by using low angle or undertree sprinklers reduces the absorption problems. However, lower leaves that are moistened may be affected. Pivot irrigation sprinklers can be modified with drop lines to apply the water to the soil and not to leaves.

7. Increasing droplet size results in less absorption. However, large droplets may induce soil dispersion, sealing, and compaction which may cause greater run off.

8. In extreme cases, it may be necessary to change from the more sensitive crops such as beans and grapes.

9. Crops planted in the cooler season have a better chance to mature before the sodium or chloride can accumulate to concentrations that may cause toxicity.

Bernstein and Francois[24] stated that annual crops may be more sensitive to salts taken up through the leaf during sprinkling than to similar water salinities applied by the surface or drip methods.

Drip Irrigation

The method depends on conveying water through plastic pipes along the furrows. At each plant there is a dripper from which drops of water moisten the soil surrounding the plant. The drippers are adjusted to supply the plant with 2–10 l per h. The water moistens the soil and penetrates through it in the form of a cone or an onion shape. The distance between the drippers is adjusted so that the moistened soil circular surfaces touch each other. The distance between the pipe lines is adjusted according to the method of planting. One pipe line may suffice two rows of plants growing at narrow spaces, while trees in orchards may need 2 or more pipe lines per one row of trees. It suits any topographic condition suitable for row crop farming (FAO[18]).

Clogging of Drip Irrigation System

Clogging of the emitters of the drip irrigation system constitutes an important problem, especially when the emitters are partially clogged.

Tables 32 and 33 (Bucks et al.[27]) show the principal physical, chemical, and biological contributors to clogging. Ayers and Westcot[23] recommended that a complete water analysis be conducted before a system is designed in order to allow for treatment to improve water quality before it reaches the small openings. Nakayamaz,[28] as stated by Ayers and Westcot,[23] reported the information shown in Table 34 which presents a relative scale for situations when clogging problems may occur due to water quality.

Causes of Clogging

Solid particles in suspension

They usually consist of soil particles, lime carbonates, solid material washed into canals, algae, and eroded material from reservoirs.

Filtration is more reliable than sedimentation. Various screening materials and filters are available. New emitter designs, some of which are self-cleaning, greatly reduce the plugging hazard.

Precipitation

Precipitation of $CaCO_3$ and phosphates can result from an excess of calcium or magnesium carbonates and sulfates. Also, iron in the ferrous form may be oxidized to insoluble ferric form.

Calcium precipitation may take place—according to the Langelier saturation index—which equals the difference between the actual pH of water (pH_a) and the theoretical pH (pH_c).

Calcium precipitation can be expected from calculating the Langelier saturation index:

$$\text{Saturation index} = pH_a - pH_c \qquad (30)$$

where pH_a is the actual pH of the water and pH_c is the theoretical pH that water could have if in equilibrium with $CaCO_3$.

When pH_a is greater than pH_c, i.e., the difference is positive, there is possibility that $CaCO_3$ precipitate would take place, whereas negative values of $pH_a - pH_c$ indicate that the water will dissolve $CaCO_3$. Table 34 shows the calculation of the value pH_c from the relationship

$$pH_c = (pK_2 - pK_c) + pC_a + p(Alk) \qquad (31)$$

Controlling the water pH or cleaning the system periodically with acid are effective means to avoid clogging due to Ca precipitation (Ayers and Westcot[23]; Nakayama[28]).

It is recommended that acidification to not lower than 6.5 is practiced periodically using hydrochloric or sulfuric acid. Sulfur burners form SO_4 which is used to acidify the supply water before entering the dripping system (Ayers and Westcot[23]).

Table 32.   Physical, chemical and biological contributors to clogging of localized (drip) irrigation systems as related to irrigation water quality.

Physical (suspended solids)	Chemical (precipitation)	Biological (bacteria and algae)
1. Sand	1. Calcium or magnesium carbonate	1. Filaments
2. Silt	2. Calcium sulphate	2. Slimes
3. Clay	3. Heavy metal hydroxides, oxides, carbonates, silicates and sulphides	3. Microbial depositions (a) Iron (b) Sulphur (c) Manganese
4. Organic matter	4. Fertilizers (a) Phosphate (b) Aqueous ammonia (c) Iron, zinc, copper, manganese	4. Bacteria
		5. Small aquatic organisms (a) Snail eggs (b) Larva

Source: Adapted from Bucks et al.[27] (1979).

Table 33.   Influence of water quality on the potential for clogging problems in localized (drip) irrigation systems.

Potential Problems	Units	Degree of Restriction on Use		
		None	Slight to Moderate	Severe
Physical				
Suspended solids	mg/l	<50	50–100	>100
Chemical				
pH	mg/l	<7.0	7.0–8.0	>8.0
Dissolved solids	mg/l	<500	500–2000	>2000
Manganese[1]	mg/l	<0.1	0.1–1.5	>1.5
Iron[2]	mg/l	<0.1	0.1–1.5	>1.5
Hydrogen sulphide	mg/l	<0.5	0.5–2.0	>2.0
Biological				
Bacterial populations	maximum number/ml	<10,000	10,000–50,000	>50,000

[1]   While restrictions in use of localized (drip) irrigation systems may not occur at these manganese concentrations, plant toxicities may occur at lower concentrations (see Chapter 3, Table 21).

[2]   Iron concentration >5.0 mg/l may cause nutritional imbalances in certain crops (see Chapter 3, Table 21).

Source: Adapted from Nakayama[28] (1982).

Table 34.   Procedure for calculation of pHc.[1]

pHc = (pK$_2$ - pKc) + pCa + p(Alk)

pK$_2$ – pKc is obtained from the concentration of Ca + Mg +   Obtained from the
  Na in me/l   water analysis

pCa is obtained from the Ca in me/l

p(Alk) is obtained from the concentration of CO$_3$ + HCO$_3$ in
  me/l

Concentration (me/l)	pK$_2$ – pKc	pCa	p(Alk)
0.5	2.0	4.6	4.3
0.10	2.0	4.3	4.0
0.15	2.0	4.1	3.8
0.20	2.0	4.0	3.7
0.25	2.0	3.9	3.6
0.30	2.0	3.8	3.5
0.40	2.0	3.7	3.4
0.50	2.1	3.6	3.3
0.75	2.1	3.4	3.1
1.00	2.1	3.3	3.0
1.25	2.1	3.2	2.9
1.50	2.1	3.1	2.8
2.00	2.2	3.0	2.7
2.50	2.2	2.9	2.6
3.00	2.2	2.8	2.5
4.00	2.2	2.7	2.4
5.00	2.2	2.6	2.3
6.00	2.2	2.5	2.2
8.00	2.3	2.4	2.1
10.00	2.3	2.3	2.0
12.50	2.3	2.2	1.9
15.00	2.3	2.1	1.8
20.00	2.4	2.0	1.7
30.00	2.4	1.8	1.5
50.00	2.5	1.6	1.3
80.00	2.5	1.4	1.1

[1]   pHc is a theoretical, calculated pH of the irrigation water.
Source: Procedure from Nakayama[28] (1982).

Clogging potential due to precipitation of iron is somewhat difficult because iron is involved in other clogging problems. Table 33 shows that microbial deposition which causes clogging of the emitters is affected by iron, sulfur, and manganese. Presence of 5 mg/l is the maximum concentration limit of iron in drip irrigation (Table 35). However, in practice this maximum should not be above 2.0. According to Ayers and Westcot,[23] 0.5 mg/l constitutes a potential problem if tannin-like compounds (often in acid waters) or total sulfides exceed 2 mg/l. The combination of the two normally produces undesirable slime growths.

To avoid iron clogging of lines or emitters, iron should be oxidized by chlorination, precipitated, and filtered out before it enters the irrigation system. Oxidation by air injection is also practiced to precipitate iron (Ayers and Westcot[23]).

Precipitation may also take place if fertilizers are injected into the irrigation, a practice known as "fertigation." Presence of Ca in the water in a concentration greater than 6 meq/l

Table 35.   Recommended maximum concentration of trace elements in irrigation waters.

Element (symbol)	For Waters Used Continuously on all Soils mg/l	For Use Up to 20 Years on Fine-Textured Soils of pH 6.0 to 8.5 (mg/l)
Aluminum (Al)	5.0	20.0
Arsenic (As)	0.1	2.0
Beryllium (Be)	0.1	0.5
Boron (B)	1/	2.0
Cadmium (Cd)	0.01	0.05
Chromium (Cr)	0.1	1.0
Cobalt (Co)	0.05	5.0
Copper (Cu)	0.2	5.0
Fluoride (F)	1.0	15.0
Iron (Fe)	5.0	20.0
Lead (Pb)	5.0	10.0
Lithium (Li)[2/]	2.5	2.5
Manganese (Mn)	0.2	10.0
Molybdenum (Mo)	0.01	0.05[3/]
Nickel (Ni)	0.2	2.0
Selenium (Se)	0.02	0.02
Vanadium (V)	0.1	1.0
Zinc (Zn)	2.0	10.0

These levels will normally not adversely affect plants or soils. No data available for Mercury (Hg), Silver (Ag), Tin (Sn), Titanium (Ti), Tungsten (W). *Source:* Environmental Studies Board, Natl. Acad. of Sci., Natl. Acad. of Engineering.[60]

causes the precipitation of P fertilizer, thus clogging of emitters. Anhydrous or liquid ammonia if injected into the water raises the water pH to about 11, thus precipitating $CaCO_3$ and causing clogging of the entire system.

It has been stated above that biological activities by microorganisms such as algae, slimes, fungi, bacteria, or snails and larvae cause clogging of the irrigation lines and openings. Several factors affect the growth of these biota among which are the water content of organics, iron, or hydrogen sulfide.

Chlorination is an effective treatment for controlling biological growth though it is costly and requires much care to conduct safely (Ayers and Westcot[23])

Automated Irrigation

Automated irrigation is practiced in several farms to achieve specified objectives (Montanes[29]):

1. Saving irrigation water. This objective protects water from loss and eliminates the application of excessive amounts of water which cause harmful effects to plants, soil salinization, and waterlogging.
2. Saving in irrigation labor cost.

3. Controlling the irrigation frequency according to the plant need.
4. Automatic irrigation systems guarantee water supplies, thus avoiding manual errors and ill will on the part of the operators. Since water transport and distribution are not subject to error or malice, a climate of confidence prevails among the farmers. Automation may be designed to meet particular purposes:

    a. The whole irrigation operation may be automated.
    b. In free level canals, automated gates respond adequately to fluctuations in consumption demand.

In automated irrigation two approaches may be adopted (Montanes[29]):

a. Irrigation "on demand," i.e., the farmer himself decides the time to irrigate.
b. The "preestablished system" described by Montanes[29] as follows: "When a zone reaches the moment for irrigation to start, it is triggered off by a single manual operation and continues in accordance with preset programs for the various plots, automatically followed by hydraulic, mechanical or electrical systems."

Differences between irrigation on demand and preestablished irrigation are:

•   On demand users may wish to irrigate at a specified time, thus the network has to convey flows greater than the average continuous flow while with preestablished irrigation, the maximum flow can be decreased which subsequently decreases the distribution network's discharge capacity.
•   On demand system is more suitable for irrigation systems with semi-fixed irrigation installations. The preestablished system is much less suitable for manual operation. Initiation of the irrigation operation would be controlled by an order from the tensiometers or other devices indicating the need for irrigation.

Automated irrigation has been achieving significant progress, especially after electronic computers had become common place. This is especially true in regions which need more water than what they have got. Thus, efficient use of water is among their main concern. However, automated irrigation requires a special scientific standard as well as awareness of the importance of water and the advantages of automating irrigation. The following is a short description of an automated irrigation system (Arlosoron[30]).

The system is remote controlled including follow-through systems and is capable of performing the following operations:

•   automatic follow-through switching of control groups according to water quantities and the order of activation, as determined at the control panel.
•   the opening and closing, at will, of values in the field.
•   feedback.
•   the possibility of connecting various sensors to the control panel, an electronic computer, a channel for a regional computer, and any other additional control accessories.

Technologies for Saving Water

In arid ecosystems, water is usually a limiting factor for land utilization. The growers in the arid regions should know that the area which they may utilize depends on the amount of water they have. The amount of water utilized per hectare may be decreased if specified principles and technologies are adopted such as:

- The farmer should be persuaded that excess water is as harmful as drought. Under excess water, plants suffocate, they cannot uptake nutrients due to lack of oxygen and their growth ceases. Irrigation by gravity without lifting encourages the farmers to apply excess water. Sprinkler and drip irrigation methods are successful means to save water.

  If water is a limiting factor, the fine-textured soils require less water than the coarse-textured soils.
- Selection of crops is another means for saving water. Barley water consumption is less than clover and rice requires much more water than maize.
- Technologies to control losses of water by seepage and evaporation from the canals or the fields stated in the utilization of the sandy soils (Chapter 4) may be applied.
- Technologies to decrease loss of water by evaporation are recommended starting with mulching using plant residues, plastic, and asphalt. Evaporation of water from water bodies may be decreased with chemicals such as saturated alcohols. In warm arid regions, it may be recommended to leave the land fallow in summer, thus avoiding the high evapotranspiration in this season.
- Method, amount, and timing of irrigation are means for improving water efficiency.

The role of evaporation in water loss may be decreased by growing plants that fully cover the field (Loomis[31]). He also is in favor of applying the leaching requirements occasionally rather than at each irrigation. He suggests applying only some fraction of weather-based estimates of ET_o, or lengthening the irrigation cycles, thus restricting water use by plants through mild stress provided the crops can tolerate this "deficit irrigation" without greatly affecting its production.

- Technologies to reduce transpiration by spraying have also been practiced.
- Chemicals coating the leaves such as by kaolinite spray to modify the net radiation balance. However, kaolinite has broad reflectance over the whole short wave spectrum, Loomis.[31] Loomis suggested selecting reflection of the high-energy photons of blue light (good quantum efficiency in photosynthesis but poor energy efficiency).
- Spraying with antitranspirant materials which either add an additional diffusion barrier over the stomates such as polythelene or chemically regulate stomatal resistance with phenyl mercuric acetate and growth regulators such as abscisic acid which can induce stomatal closure, Loomis.[31]
- Other technologies that reduce transpiration are those related to modifying the aerial environment above the plant which may modify radiation or increase aerodynamic resistance to water loss.
- Formation of cloud over the crop (e.g., by jet contrail development).
- Glass or plastic houses.
- Screen shelters.
- Windbreak shelters.

Calculation of Irrigation Requirement of Water

1. Potential evapotranspiration "ET" calculated from Penman Equation using the climatic records (see Chapter 2, Table 17).
2. "ET" crop factor using the factors published by the FAO/UNESCO, cited in Ayers and Westcot.[23]

3.   The conveying efficiency for the method of irrigation canals (the value of 0.9 may be used when the canals are lined), thus the efficiency of the project irrigation can be known.
4.   The value of "ET" is divided by the project irrigation efficiency results in the water requirement for the crop.
5.   The leaching requirement is calculated and added to the crop water requirement.

FAO[18] describes a simple method to determine depth of water to be applied and frequency of its application as follows.

Assuming that the rooting depth of the crop (e.g., potato) is 55 cm, irrigation is required when the soil water tension reaches 0.5 bar under evapotranspiration rate of 5 mm/day. The depth and frequency of irrigation will be:

Depth of water to be applied:

$$\text{Summation of } \frac{(b-c) \times d \times a}{100} \text{ for horizon 1 and 2} = 2.5 \text{ cm}$$

$$\text{Irrigation interval in days} = \frac{\text{depth of irrigation}}{\text{Evapotranspiration}} = \frac{2.5}{0.5} = 5 \text{ days}$$

It should be pointed out that the depth of water calculated is only the consumptive use. Irrigation efficiency and leaching requirement should be taken into consideration. The evapotranspiration has to be adjusted using the crop coefficients which indicate the crop water requirements at different stages of growth.

Efficiency of Water Conveyance

It is the ratio between the amount of water which reaches the field and the amount discharged into the canal at its terminal, e.g., if the amount of water entering the field is 2.0 m³/min, and the amount discharged into the canal is 2.5 m³/min, thus the efficiency of water conveyance is $2.0/2.5 \times 100 = 80$ (see Chapter 3) (Land Master Plan[32]).

Loss of water in the conveying canal may be due to:

1.   Infiltration from the sides and/or the bottom of the canal.
2.   Evaporation from the water surface.
3.   Evapotranspiration via weeds or trees on the banks of the canal.
4.   Cracks in the canal banks or in masonry built on the canal.
5.   Water weeds which clog the waterway causing water spill on both canal sides.

DRAINAGE

It has been clearly understood that irrigation projects in arid regions are hazardous unless accompanied with efficient drainage. Without drainage, the water table approaches the soil surface, thus decreasing the soil air, limiting the root system, and lowering the soil productivity. When the groundwater approaches the soil surface by capillarity, it evaporates leaving its content of salts to accumulate by time. Under such conditions, plants suffer from salinity as stated in discussing salt-affected soils (in Chapter 2). Waterlogging has several other effects on the soil microbiology and on the soil chemical reactions. This is why drainage which protects the soil from waterlogging and subsequent salinization is a compulsory process wherever irrigation is introduced and especially in arid and semiarid regions.

Role of Groundwater in Soil Salinization

In regions where groundwater lies at shallow depths, water moves upward to the soil surface. The salt content of this water accumulates at or near the soil surface. The upward movement of groundwater and its subsequent evaporation has received considerable attention from several investigators. Gardner[33] reviewed the studies on water movement in the soil profile from the thermodynamic aspects. Doering et al.[34] studied the distribution of salts accumulated by evaporation in the soil profile. Willis[35] applied mathematical relationships to describe the effect of layers different in texture on the rate of evaporation from soil columns. Balba and Soliman[36] attempted to estimate the role of both liquid and evaporated water which moved upward from a constant water level, in salinization of homogeneous and layered soil columns differing in texture. They concluded that the soil texture affected the retained liquid water, the evaporated water, and the rate of evaporation. Salinization of the soil columns due to evaporation increased with time and varied according to the soil texture. When soil moisture content of the soil column reaches the equilibrium state it does not change with time (Figure 16).

Again Balba and Soliman[37] studied the effects of cations, anions, and their concentrations on the accumulation of salts by evaporation. They concluded (Figures 17–19):

- The presence of salts in the shallow water table resulted in a decrease in the evaporated amount of water. Increasing salt concentration of water caused a corresponding decrease in evaporated water. The evaporated amount of water varied with the kind of cation in the following order: $MgCl_2 < CaCl_2 < NaCl <$ KCl (Balba and Soliman[37]).
- Differences in evaporated amount of water due to different anions were not pronounced as in the case of cations (Balba and Soliman[37]).

The highest level to which the water table should be permitted to rise during an irrigation season depends on several factors (FAO[18]):

- The soil capillarity characteristics and its aeration.
- The climatic conditions such as the prevailing evaporation rate, intensity, amount, and frequency of rainfall.
- The crop characteristics such as depth of rooting, tolerance to salts and to the water table.
- Quality of irrigation as well as of groundwater.

As a general rule, a higher level of water table is tolerable in a cool area than in a hot dry area, particularly if the water in the hot area is saline.

Though it is advisable to maintain the water table depth at 300 cm from the soil surface, 90–120 cm depth is considered satisfactory for several crops.

El Gabaly and Naguib[38] studied the level and salt concentration of groundwater with regard to their effect on the salinization process using lysimeters. They concluded that:

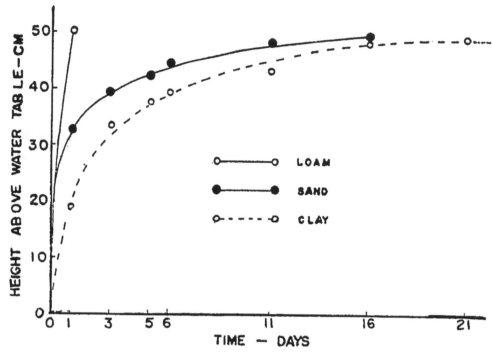

Figure 16. Upward movement of groundwater by time in soils different in texture (Balba and Soliman[36]).

- When the water table was kept at a depth of 50 cm, the increase in salinity in the upper 20 cm was very pronounced. At a depth of 90 cm the soil salinity was less than 1/3 of its level at a depth of 50 cm.
- The depth of groundwater contributes more to the salinity of the soil surface than does the salinity level of the groundwater.
- The concentration of chloride and sodium in the top 20 cm layer was higher than in the next 20 cm, particularly during the fallow period and at a 50 cm depth of groundwater.

The investigations to be conducted before planning the irrigation system stated above include studies pertaining to drainage planning such as topography, slope, hydraulic conductivity, presence and depth of impermeable or slowly permeable soils horizons, etc. These investigations and others should be carried out before setting the drainage scheme.

The drainage system may be:

a.   A network of open ditches discharging their water into a collector which in turn discharges its water either into a larger public collector or into an outlet which may be a lake or the sea or merely a depression.

b.   The above network of ditches might be reached by underground tiles perpendicular to the land slope. The parcel tiles discharge the water they receive into a collector underground tile, or into an aboveground ditch. The drained water should be discharged into an outlet.

c.   Vertical drainage: Wells dug to reach the underground drained water. Water is pumped from these wells and may be reused for irrigation provided its quality is suitable. With pumped drainage (tube wells) it is advisable to hold the water table below a depth of 300–400 cm as a safety precaution in the event of a well failure.

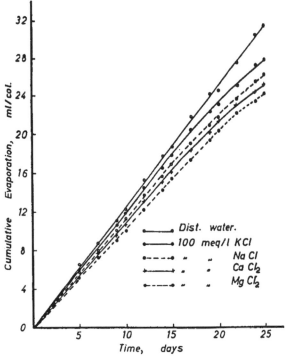

Figure 17. Cumulative evaporation vs. time from shallow water table containing different chloride salts (Balba and Soliman).

The quality of groundwater usually tends to deteriorate over time (FAO[18]).

Reuse of the drainage water collected by a drainage system for irrigation is permissible depending on its salinity level and composition as will be discussed in detail. The water quality, however, should be monitored.

d. Dead drains: Drains without outlets. They depend on the evaporation of the water they receive.

e. Biological drainage: Growing trees along the collector drain. Thus the evapo-transpiration of water via the trees reduces the water level of the collector drain to a level lower than the levels in the ditches discharging in the collector.

f. Mole drains: Voids made underground by passing a bomb-shape cone 30–50 cm long and 5–8 cm in diameter which forms a tunnel 60–90 cm deep. This kind of drain suits clay soils with low hydraulic conductivity.

USE OF LOW-QUALITY WATER FOR IRRIGATION

Lack of water is an obstacle confronting development in countries of the arid and semiarid regions. Water sources of limited quality such as drainage and underground waters were suggested for agricultural use. In order to assess the suitability of water for irrigation, several criteria should be studied and certain practices should be followed in its use.

Schemes of water classification according to its suitability for irrigation were established, but it was always felt that they were not satisfactory. A review of work on irrigation water quality was reported by Richards.[13] Modifications of the classification of Wilcox[39] and Thorne and Thorne[40] were adopted in the U.S. Salinity Laboratory class-

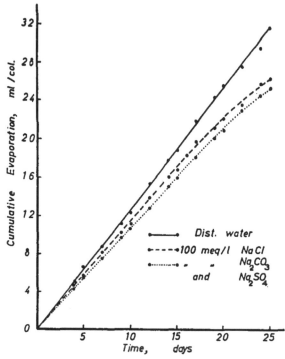

Figure 18. Cumulative evaporation vs. time from shallow water table containing different sodium salts (Balba and Soliman).

ification (Richards[13]). Rhoades[41] pointed out that salinity hazards of a water should be judged through the potential of this salinity to create conditions hazardous to plant growth. He stated that the total salt content of irrigation water is only of a general descriptive value and the limits established in one system of classification are not applicable without restriction to every place and condition. Ayers[42] related the degree of problems that may affect the plants to the salt concentration of irrigation water.

Mechanism of Soil Salinization with Saline Irrigation Water

When water is added to the soil, several processes take place: cation exchange reaction, dissolution or precipitation of salts, and accumulation in, or leaching from the soil, of soluble salts with the water movement through the soil. The U.S. Salinity Laboratory Staff,[13] Bryssine[43] cited by Durand and several other investigators such as Gardner and Brooks,[44] Brooks et al.,[45] and Balba and Bassiuni[46] attempted to calculate the salt concentration in the soil layers after irrigation.

Balba[47] studied the mechanism of salt accumulation and removal in glass columns using soils differing in their salinity level and texture leached with water having different $CaCl^2$ concentrations. The study showed the following:

- The amount of salt removed from the soil column per kg soil increased with the increase of the initial soil salt content.
- The fine-textured soils lost less amounts of their salt than the coarse-textured soils.

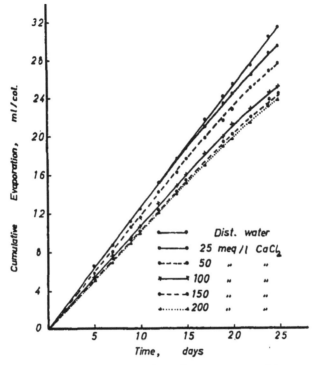

Figure 19. Cumulative evaporation vs. time from water table containing different concentrations of CaCl$_2$ (Balba and Soliman).

- When the soil column was leached with different amounts of saline water the final salt content in the soil column did not materially vary with the variation of the added amount of water.
- Also, it was found that the final salt content after one irrigation was almost the same after 2 or more irrigations.
- The clay soils retain greater amounts of water. Hence the amount of salt retained from the saline water was greater than the corresponding amounts retained by coarse-textured soils.
- Use of ^{22}Na showed that the ability of the applied solution to carry the soil salt in its passage through the soil column decreases with an increase in the salt concentration of the applied water. This might be explained as being due to a decrease in the water's ability to dissolve salt with an increase in its original content of salts, Figure 20; Balba and Bassiuni.[46]

The potential salinization of soils irrigated with saline waters depends not only on the water salinity level but also on several other conditions such as the following:

- The irrigation regime, especially the period between irrigations.
- The amount of applied water.
- The method of irrigation.
- The depth of groundwater table.
- The hydraulic conductivity.
- The efficiency of the drainage system.
- The climatic conditions.
- The soil texture.

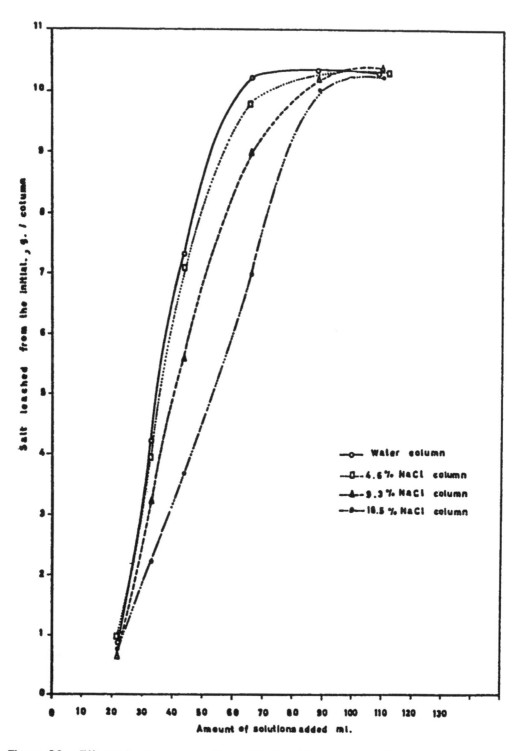

Figure 20. Effect of salt concentration of the leaching water on the amount of leached salt from sand columns (Balba and Bassiuni).

- The presence of impermeable layers and their depth.
- The kind of vegetative cover.
- The soil topography.

Water Quality Classification*

In 1954, the U.S. Salinity Laboratory (Richards[13]) classified water suitability for irrigation from its salinity standpoint into the following classes:

Low: 0.1–0.25, medium: 0.25–0.750, high: 0.75–2.250, and very high: above 2.250 dS m^{-1}. This classification enjoyed wide popularity among water laboratories though the "EC" of the tap water in many communities is classified as "medium" according to this scheme.

Ayers and Westcot[23] considered that "water suitability is determined by its potential to cause problems and is related to the special management practices needed or the yield reduction caused." Accordingly, they looked at the water quality for irrigation from the angle of the degree of the problems that may occur as a result of utilization of the water. The problems that may be caused as a result of water use are salinity, low infiltration rate, specific ion toxicity, and others. They introduced a water classification—guidelines—in 1976 and slightly modified it in 1985 based on the degree of restriction on use of the water resulting from the potential irrigation problems: salinity, slow infiltration, specific ion toxicity or miscellaneous effects (Table 36).

Rating water for irrigation should take the sodicity hazards into consideration. Eaton[48] used two terms for this evaluation:

1. Soluble sodium percentage possible = (Na × 100/Ca+Mg+Na).
2. Residual (Na$_2$CO$_3$ + NaHCO$_3$) = (CO$_3^-$+HCO$_3^-$) – (Ca^{++}+Mg^{++}) where the ionic constituents are expressed in meq per liter.
3. The U.S. Salinity Laboratory (Richards[13]) used the ratio SAR = Na/(Ca+Mg)/2 as an index of sodicity hazard of irrigation water.

When the soil is equilibrated with water containing Na, Ca, and Mg, represented by "SAR" value, the exchangeable sodium on the soil surface was found to differ according to the anion composition of the applied water. The adsorbed sodium when CO$_3^-$ or OH$^-$ prevail is greater than in the case of SO$_4^-$ or Cl$^-$. This was shown by Balba and Balba[49] to be due to (1) increase of the cation exchange capacity of the clay in high pH values, and (2) precipitation of displaced Ca^{++}.

In the case of SO$_4^-$-water, the adsorbed Na was greater than in case of Cl$^-$-water. The difference was shown to be due to (1) differences in the activity coefficient of cations in the equilibrium solution, and (2) formation of ion pairs between cations especially Ca^{++} and anions especially SO$_4$ (Talur et al.[50]; Singh and Turner[51]; Balba and Balba[49]). Accordingly, sulfatic water is less suitable for irrigation than Cl$^-$-water as it will aggravate the infiltration problem due to excess ESP.

Management of Toxicity Problems

Changing the water source may be the ultimate method to get rid of the water toxicity problem. However, this change may not be possible.

---

*   The effect of salinity and sodicity on soils and plants was discussed in Chapter 2 of this book.

Table 36. Guidelines for interpretations of water quality for irrigation.[1]

Potential Irrigation Problems	Units	Degree of Restriction on Use		
		None	Slight to Moderate	Severe
Salinity (affects crop water availability)[2]				
EC_w	dS/m	<0.7	0.7–3.0	>3.0
TDS	mg/l	<450	450–2000	>2000
Infiltration (affects infiltration rate of water into the soil. Evaluate using EC_w and SAR together)[3]				
SAR = 0–3 and EC_w =		>7.0	0.7–0.2	<0.2
= 3–6 =		>1.2	1.2–0.3	<0.3
= 6–12 =		>1.9	1.9–0.5	<0.5
= 12–20 =		>2.9	2.9–1.3	<1.3
= 20–40 =		>5.0	5.0–2.9	<2.9
Specific Ion Toxicity (affects sensitive crops)				
Sodium (Na)[4]				
surface irrigation	SAR	<3	3–9	>9
sprinkler irrigation	me/l	<3	>3	
Chloride (Cl)[4]				
surface irrigation	me/l	<4	4–10	>10
sprinkler irrigation	me/l	<3	>3	
Boron (B)[5]	mg/l	<0.7	0.7–3.0	>3.0
Trace Elements (see Table 21)				
Miscellaneous Effects (affects susceptible crops)				
Nitrogen (NO_3 - N)[6]	mg/l	<5	5–30	>30
Bicarbonate (HCO_3) (overhead sprinkling only)	me/l	<1.5	1.5–8.5	>8.5
pH		Normal Range 6.5–8.4		

[1] Adapted from University of California Committee of Consultants, 1974.

[2] EC_w means electrical conductivity, a measure of the water salinity, reported in deci-Siemens per meter at 25°C (dS/m) or in units millimhos per centimeter (mmho/cm). Both are equivalent. TDS means total dissolved solids, reported in milligrams per liter (mg/l).

[3] SAR means sodium adsorption ratio. SAR is sometimes reported by the symbol RNa. See Figure 1 for the SAR calculation procedure. At a given SAR, infiltration rate increases as water salinity increases. Evaluate the potential infiltration problem by SAR as modified by EC_w. Adapted from Rhoades, 1977, and Oster and Schroer, 1979.

[4] For surface irrigation, most tree crops and woody plants are sensitive to sodium and chloride; use the values shown. Most annual crops are not sensitive; use the salinity tolerance tables (Tables 4 and 5). For chloride tolerance of selected fruit crops, see Table 14. With overhead sprinkler irrigation and low humidity (<30%), sodium and chloride may be absorbed through the leaves of sensitive crops. For crop sensitivity to absorption, see Tables 18, 19 and 20.

[5] For boron tolerance, see Tables 16 and 17.

[6] NO_3-N means nitrate nitrogen reported in terms of elemental nitrogen (NH_4-N and Organic-N should be included when wastewater is being tested).

Source: Ayers and Westcot.[23]

Leaching would be the second thought. This process does not solve the problem as it is renewed with each irrigation. In this regard boron is difficult to leach as it is adsorbed as BO_3^- on the soil colloidal complex.

Application of calcium to the applied water or to the soil to be irrigated with Na-rich water is considered an effective technique to counteract Na toxicity.

Changing the crops to others more tolerant to boron or chloride toxicity can also be a solution for the problem though it is not always feasible.

An alternative water supply may also help in controlling the degree of toxicity at bearable levels. In this regard, blending fresh and poor water to minimize the toxicity hazard of the poorer water may be suggested if feasible.

As stated above, sprinkler irrigation should be changed if water content of Na or Cl is above 3 meq/l (Ayers and Westcot[23]).

Practices of Saline Water Utilization

Use of saline waters for irrigation requires certain precautions which should be satisfied, otherwise the soil will be deteriorated and crops yields will decrease. However, even with these precautions and practices, soils irrigated with saline waters will be salinized to a certain degree depending on the salinity of the water used for irrigation, provided the appropriate leaching requirements are applied and an efficient drainage system is present. The average expected soil salinity of the soil paste extract "ECe" after being irrigated with saline water is approximately evaluated from ECe = 3/2 EC_{iw} (Ayers[42]).

The Leaching Requirement

Crop production is reduced when excessive accumulations of soluble salts exist in soils. This accumulation of salts occurs with irrigation, even when the original soil is low in salt, because plants absorb and transpire water from the soil, while the salt in the irrigation water is left behind in the soil. This salt frequently accumulates in the soil to the point that plant growth is reduced unless the salt is removed by leaching. Irrigation with sufficient water in excess of the evapotranspiration requirement will achieve the needed leaching which in turn requires adequate drainage. The leaching requirement "LR" is defined as the fraction of the irrigation water that must be leached through the root zone to keep the salinity of the soil below a specified value (Richards[13]).

From the salt balance in the soil upon irrigation,

$$V_{iw} \cdot C_{iw} + S_m - V_{dw} - S_p - S_c = O \tag{32}$$

where V_{iw} and V_{dw} are volumes of irrigation and drainage waters, respectively; C_{iw} and C_{dw} are the corresponding concentrations of both waters; S_m, S_p, and S_c are the amounts of salts dissolved from soil minerals, precipitated in soil, and removed by the harvested crops.

Disregarding the values of S_m, S_p, and S_c, the equation reduces to:

$$V_{iw} \cdot C_{iw} = V_{dw} \cdot Cdw \tag{33}$$

Replacing the depth "D" and the electrical conductivity "EC" for the volume and the concentration, respectively, and rearranging, Equation 33 becomes:

$$EC_{iw} / EC_{dw} = D_{dw} / D_{iw} = LF \qquad (34)$$

Equation 34 shows the equality between the leaching fraction $(LF)_7$ $(D_{dw}/D_{iw}$, the ratio between the drained water depth and irrigation water depth), and EC_{iw}/EC_{dw}.

When a maximum permissible "EC" value based on crop tolerance to salinity is substituted for "EC_{dw}" in Equation 34, the ratio "D_{dw}/D_{iw}" is the extra amount of irrigation water that must be passed through the root zone to prevent the "EC_{dw}" from exceeding the permissible level.

To evaluate the leaching requirement "LR," the salinity of irrigation water "EC_{iw}" and the crop tolerance to soil salinity "ECe," must be known. Uniform root zone soil extract "ECe" values that produce 50% yield decrease in forage, field, and vegetable crops and 10% yield decrease in fruit crops have been used as reasonable estimates for "EC_{dw}" in Equation 34 (Bernstein[52,53] and Bower et al.[54]). The basis of this approach lies on the fact that for a uniform textural soil, the "EC_{sw}" values under steady state conditions increase with depth in the profile and the "ECe" attains the value at which crop yield is one half that attained under nonsaline conditions. Thus the "EC" of the soil water below the root zone can exceed a value of EC_{50} SE without appreciable salinity build-up in the root zone.

Bernstein and Francois[24] concluded from extensive experiments that crop growth is relatively insensitive to high salinities in lower root zone regions and that leaching requirements can be reduced to one fourth the levels calculated from Equation 34. They suggested that the conventional "ECe" values be increased 4-fold before substitution in Equation 34 for "EC_{dw}."

An alternative procedure, suggested by Rhoades[55] to select appropriate "EC" value, may be derived from observations that (1) the average "ECe" in the root zone is related to the "ECe" values found at the top "EC_t" and bottom "EC_b" of the profile as follows:

$$\text{Average } EC_{se} = K \frac{(EC_t + EC_b)}{2} \qquad (35)$$

where "K" is about 0.8 at relatively low leaching fractions and (2) "EC_t" and "EC_b" for saturation extract values are approximately equal to the product Q_{Fe}/Q_{Se} (soil moisture content at field capacity and at saturation percentage) times "EC_{iw}" and "EC_{dw}," respectively. With substitution of the above relations into Equation 35 with Q_{se}/Q_{Fe}^{-2} and making the assumption that crops respond to average root zone salinity, the following equation was obtained and can be used to calculate appropriate "EC_{dw}" values:

$$EC_{dw} = 5 EC_{Se} - EC_{iw} \qquad (36)$$

where "EC_{Se}" is the average "EC" of saturation extract for a given crop appropriate to the tolerable degree of yield depression (usually 10% or less).

This procedure makes the choice of "EC_{dw}" for the denominator of Equation 34 used to estimate the "LR" less empirical than the original (historical) and additionally makes the selection process also dependent upon the concentration of the irrigation water.

The value of "LR" according to the three methods presented above—the historical,[13] Bernstein and Francois,[25] and Rhoades[56]—for a specified water differs according to the method of calculation (see Chapter 3, Table 20).

The Leaching Requirement Concept for SAR Control

Rhoades[56] suggested that the maximum permissible "SAR" value would be that SAR which corresponds to the ESP (exchangeable sodium percentage) at which the crop grown would suffer from sodium toxicity, or the "ESP" at which soil permeability would be reduced. Rhoades' idea in applying leaching requirements for SAR control is based on the observation that ESP decreases with the increase in the applied water. To estimate "LR_{SAR}" for soils, Rhoades substituted the appropriate values for "SAR_{sw}," "SAR_{iw}," and "PH_c" in the following equation:

$$SAR_{sw} = SAR_{iw} (1 + 8.4 + PH_c) \tag{37}$$

where "SAR_{sw}" is the chosen "SAR" value and "R" is a constant. The value of "K" is thus known. Rhoades[34] established Figure 21 from which "LR_{SAR}" which corresponds to the obtained "K" value is known.

Rhoades[55] suggested the following equation to estimate the maximum allowable "EC_{iw}" termed "EC_{iw}" for irrigation water under conditions of average irrigation frequencies:

$$EC_{iw} = s \text{ av. } EC_{10} / 1 + \frac{1}{LF} \tag{38}$$

where "ECe_{10}" is the average "ECe" of the root zone saturation extract for a given crop appropriate to a negligible reduction of about 10%. The best estimate for "LF" would be that obtained from estimated "EC_{dw}" from Rhoades' Equation 36.

The values in Table 37 were calculated from Equation 38. The "LF" values from 0.1 to 1.0 every 0.1 interval and "ECe_{10}" values are "EC" values of saturation extract of 2.5, 4, 5, 8, and 10 dS m^{-1} which cause 10% reduction in yield are substituted in Equation 38. For each ECe_{10} stated above, several values of "EC_{iw}" were obtained according to the value of "LF." The obtained values of "EC_{iw}" are presented in Table 37.

Timing of Leaching

In the opinion of Ayers and Westcot,[23] timing of leaching is not a critical matter provided crop tolerance is not exceeded for extended periods. They suggest that leaching can be done each irrigation, each alternative irrigation, seasonally, or even at longer intervals. Rainfall should be considered as a part of the leaching requirement as long as it infiltrates into the soil. In years of low rainfall, the surface soil layer will benefit from this rainfall and will enhance germination.

Recommended practices (Ayers and Westcot[23]) are:

- Leaching during the cool season is more effective than in the warm season as the losses of "ET" are lower.
- Salt-tolerant crops require lower "LR."
- Tillage improves water infiltration in the soil.
- Application rate of sprinkler irrigation lower than soil infiltration rate favors unsaturated flow which is more efficient than saturated flow for leaching.
- Leaching at periods of low crop water use or after the cropping season is more efficient.
- Fallow periods especially during summer should be avoided.

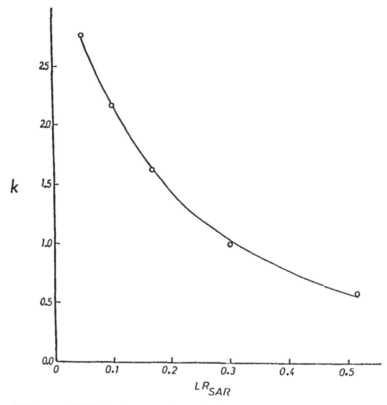

Figure 21. Values of "K" for given values of LR_{SAR} (Rhoades).

Selection of Tolerant Crops

When irrigation water is saline, one has to select crops that can tolerate salinity as it was stated in salt-affected soils (Chapter 2).

The level of salinity the crop can tolerate is a base in evaluating the leaching requirement. Maas and Hoffman[57] developed lists of field, vegetable, forage, and fruit crops with the ECe, EC_w at which percentages of their potential yields are expected. Also Maas[58] presented lists of degrees of salt tolerance of agricultural crops.

Irrigation with saline water requires application of an addition of water over and above the amount to be applied to avoid salt accumulation in the root zone. As with water salinity, waters containing toxic ions require leaching as in the case of presence of the Cl⁻ in the water. It is possible to calculate the approximate Cl⁻ concentration in the root zone after irrigating the land with the Cl⁻ water by multiplying the concentration factor, Table 36, for a certain leaching fraction by the concentration of chloride ion in the water.

A leaching requirement for Cl⁻ (LR_{cl}) was suggested by Rhoades[55] as follows:

$$LR_{(cl)} = \frac{Cl_w}{5Cl_e - Cl_w} \tag{39}$$

Table 37.   Maximum allowable EC of irrigation water and the corresponding "LF" at different ECe_{10} values.

LF	Maximum Permissible EC of Irrigation Water for Crops Tolerant to ECe_{10} dS m^{-1}				
	2.5	4	5	8	10
0.1	1.2	1.8	2.3	3.6	4.5
0.2	2.1	3.4	4.1	6.8	8.3
0.3	2.8	4.9	5.6	8.9	11.2
0.4	3.8	6.1	7.7	12.3	15.4
0.5	4.1	6.6	8.3	13.3	16.6
0.6	4.7	7.5	9.4	15.0	18.8
0.7	5.1	8.1	10.2	16.3	20.4
0.8	5.6	8.9	11.2	17.2	22.4
0.9	5.9	9.5	11.9	19.0	23.8
1.0	6.2	10.0	12.5	20.0	25.0

Source: Rhoades.[56]

where $LR_{(cl)}$ =   the minimum leaching requirement needed to control Cl with ordinary surface methods of irrigation.

Cl_w =   Cl concentration in the applied irrigation water in meq per liter.

Cl_e =   Cl concentration tolerated by crop as determined in the soil saturation extract, in meq/l.

The presence of sodium also causes toxicities to sodium-sensitive crops, at a lower SAR values in the water which might cause a soil permeability problem.

It has been stated above that LR for controlling SAR may be added to the leaching requirement to control salinity. However, above SAR value of 9 in the water, the leaching requirements would be so large that further addition would cause problems with soil aeration and drainage. Application of gypsum or other Ca-supplying amendment would be a preferred solution, otherwise change the crop to another more Na-tolerant would be advisable (Ayers and Westcot[23]).

It has been stated above that boron is more difficult to leach than Cl$^-$ or Na$^+$. Studies by Hoffman[59] have shown that the depth of water required to leach high boron soil is about 3 times as much as water will be needed.

Irrigation with Waters Containing Trace Elements

Waters usually contain trace elements in minute concentrations. Their concentrations hardly exceed a few mg/l. Generally, irrigation waters are not examined for trace elements content unless the water is contaminated with waste waters.

Several trace elements are essential nutrients that plants cannot be deprived of, though they must be given in small quantities. Excessive amounts are harmful to plants and usually cause growth reduction. Use of waste waters helped researchers to define the toxic limits of these elements in irrigation waters (Natl. Acad.[60]; Pratt[61]).

Repeated application of waters containing trace elements results in accumulation of these elements in the soil. Over time, the soil becomes contaminated with trace elements, producing uneconomic yields and/or unusable products.

Studies have shown that about 85% of the applied trace elements accumulates in the soil, especially in the surface few centimeters (Evans et al.[62]).

Under normal irrigation practices, the levels of trace elements presented in Table 35 should prevent a build up which might cause yield reduction or affect the utilization of the product. However, values in Table 38 were published in 1972. Ayers and Westcot[23] recommended that the values be considered as the maximum long-term average concentration based on normal irrigation application rates.

Problems of Water Reuse

Irrigation supplementary or year round is the backbone of agriculture in the arid and semiarid regions. Expansion in land reclamation as well as intensifying the agricultural production from the cultivated land depends on water supplies to satisfy these important activities. Several developed as well as developing countries reuse the water to satisfy their need for more water for agricultural industrial and municipal activities.

Drainage water collected from open ditches or tile drains is reused in irrigation either directly or after being mixed with good quality water. Measures that should be followed for safe reuse of this water were thoroughly discussed by Ayers and Westcot[23] and others (Table 38).

Increases in fertilization and chemical control of pesticides introduced important contaminants to the drainage. Because a part of the irrigation water infiltrates through the soil to the groundwater table, contaminants of the irrigation water may reach the groundwater. Pumping the groundwater for agricultural reuse or for municipal purposes requires precautions. Concentrations of the applied fertilizers and other chemicals should be monitored periodically.

Reusing low-quality waters in agriculture, provided applying the necessary measures, allows utilization of the good quality waters where they are necessary, especially for drinking.

Because of problems of disposing of sewage and industrial waste waters, these waters were discharged into the drainage systems and/or irrigation canals. Disposal of these effluents should be in such a way that the riverbasin water quality is protected and agricultural development is not jeopardized.

The reuse of municipal waste waters requires guidelines to estimate public health hazards. The microbial constituents and their activities associated with these effluents are a problem when reusing them in agriculture.

A meeting of experts convened by WHO (World Health Organization[63]) in 1973 concluded that (Tables 39 and 40):

- Primary treatment would be sufficient to permit reuse for the irrigation of crops that are not for direct human consumption.
- Secondary treatment and most probably disinfection and filtration are necessary when the effluent will be used for irrigation of crops for direct human consumption.

Reuse of municipal waste water in agriculture may lead to microbial contamination of air, soils, and plants in the vicinity of the irrigated land. The microbial and biochemical properties of these effluents should be evaluated and compared with the public health standards, taking into consideration the crop to be irrigated and its consumption, the soil, and water. The effluent should meet these standards before its reuse.

Table 38.   Standard water quality tests needed for design and operation of localized (drip) irrigation systems.

1. Major Inorganic Systems (see Table 2)	7. Organics and Organic Matter
	8. Micro-organism
2. Hardness[1]	9. Iron
3. Suspended Solids	10. Dissolved Oxygen
4. Total Dissolved Solids (TDS)	11. Hydrogen Sulphide
5. BOD (Biological Oxygen Demand)	12. Iron Bacteria
6. COD (Chemical Oxygen Demand)	13. Sulphate Reducing Bacteria

Source: From Ayers and Westcot.[23]

Operation of Irrigation Systems (FAO[64])

Two basic requirements for efficient operation are:

- Maintaining the irrigation and drainage works.
- Trained personnel familiar with the operation system and procedure.

An operation system:

- Should begin with an inventory identifying the system components.
- Other basic technical data concerning the operation characteristics such as canal water velocities, discharges, losses in the feeder canals, conveyance networks, farm ditches and other farm losses, freeboards of canals and drains, functional properties of hydraulic structures, etc. are also essential.
- Meteorological records are necessary.

The operation procedures should include:

- Planning of water application in accordance with cropping patterns.
- Periodical reporting of the status of water in the field.
- Assessing water requirements according to the stages of plant growth.
- Collecting and processing climatological and hydraulic data in the project area.
- Planning, allotting, regulating, and delivering water along main canals, laterals, and sub-laterals down to the farm ditches according to farm requirements.
- Disposing of excess water.
- Emergency measures during droughts or in the event of flood damages.

Maintenance

Good maintenance is a prerequisite for the efficient operation of a project.

Maintenance works in an irrigation project are in four categories: routine maintenance, annual repairs, emergency repairs, and minor improvement works.

174 MANAGEMENT OF PROBLEM SOILS IN ARID ECOSYSTEMS

Table 39. Existing standards governing the use of renovated water in agriculture.

	California	Israel	South Africa	Germany
Orchards and vineyards	Primary[1] effluent; no spray irrigation; no use of dropped fruit	Secondary[2] effluent	Tertiary[3] effluent, heavily chlorinated where possible. No spray irrigation	No spray irrigation in the vicinity
Fodder fiber crops and seed crops	Primary effluent; surface or spray irrigation	Secondary effluent, but irrigation of seed crops for producing edible vegetables not permitted	Tertiary effluent	Pre-treatment with screening and settling tanks. For spray irrigation, biological treatment and chlorination
Crops for human consumption that will be processed to kill pathogens	For surface irrigation, primary effluent. For spray irrigation, disinfected secondary effluent (no more than 23 coliform organisms per 100 ml)	Vegetables for human consumption not to be irrigated with renovated waste water unless it has been properly disinfected (1000 coliform organisms per 100 ml in 80% of samples)	Tertiary effluent	Irrigation up to 4 weeks before harvesting only
Crops for human consumption in a raw state	For surface irrigation, no more than 2.2 coliform organisms per 100 ml. For spray irrigation, disinfection, filtered waste water with turbidity 10 units permitted, providing it has been treated by coagulation	Not to be irrigated with renovated waste water unless they consist of fruits that are peeled before eating		Potatoes and cereals; irrigation through flowering stage only

[1] Primary treatment of waste water refers to the settling and removal of a portion of the suspended organic and inorganic solids.
[2] Secondary treatment refers to the activated sludge process and biological filtration (trickling filtration). It may also include retention.
[3] Tertiary and advanced treatment includes several processes depending on the use of the final product but usually includes clarification, activated carbon treatment, denitrification, and ion exchange.
Source: WHO[63] (1973).

Table 40.   Treatment processes suggested by the World Health Organization for wastewater reuse.

	Irrigation			Recreation	
	Crops Not for Direct Human Consumption	Crops Eaten Cooked; Fish Culture	Crops Eaten Raw	No Contact	Contact
Health criteria (see below for explanation of symbols)	1 + 4	2 + 4 or 3 + 4	3 + 4	2	3 + 5
Primary treatment	XXX	XXX	XXX	XXX	XXX
Secondary treatment		XXX	XXX	XXX	XXX
Sand filtration or equivalent polishing methods		X	X		XXX
Disinfection		X	XXX	X	XXX

Source: WHO[63] (1973).

REFERENCES

1. Shafei, A., Lake Mariotis, its past and its future. Bul. de L'Institute de Desert Tom 2, No. 1, 1957.
2. Zaunderez, J. and C.F. Hutchinson, A review of water harvesting techniques of the Arid Southwestern U.S. and N. Mexico (Draft) Working paper for the World Banks' Sub Saharan Water Harvesting Study. Cited in Water Harvesting, Will Critchley and Klaus Siegert, ACL/MISC/17/91 FAO, Rome, 1991.
3. Garduno, M.A., Technology and Desertification, Chapter 4 in Desertification, Its Causes and Consequences, UN Conf. Desertification Secretariat, Kenya, Pergamon Press, 1977.
4. Hammad, A., Development of Agriculture and Fish Resources in Oman Sultanate (In Arabic) p., Min of Aaric. and Fish., Oman, 1981.
5. Victor, P.E., Will the Sahara be irrigated from floating icebergs? UNESCO Courier (Arabic ed.) No. 2, 1978.
6. Arab Org. for Agric. Development (AOAD), Improving wheat production in No. E. Syria. A Consultation Report, 1974.
7. Loizidis, P.A., Report on Experiments on Rain-Fed Rotation, Ministry of Agriculture, Damascus, Syria, 1970.
8. Syrian Ministry of Agric., Crop rotation experiments in rain-fed regions, 1971.
9. Balba, A.N. and M.M. El Gabaly, Soil and Ground Water Survey for Agricultural Purposes in the N.W. Coast of Egypt, University of Alexandria Press. Bull. No. 11, 1965.
10. Kampen, J. and J. Burford, Production Systems, Soil related constraints and potentials in the semiarid tropics, with special reference to India. pp. 141–165, in Priorities for Alleviating, 1980.
11. Soil-Related Constraints to Food Production in the Tropics, jointly sponsored and published by Int. Rice Inst. and N.Y. State Col. of Agric. and Life Sci., Cornell Univ., 1980.

12. U.S. Bureau of Reclamation, Land Drainage Techniques and Standards. Rec. Instructions, USBR, Denver, CO, U.S., 1964.

13. Richards, L.A. (Ed.), The diagnosis and Improvement of Saline and Alkali Soils, USDA Handbook No. 60, 1954.

14. Viehmeyer, F.J., The availability of soil moisture to plants, Results of empirical experiments with fruit trees. Soil Sci., 114:268, 1972.

15. Wadliegh, C.H., Factors affecting soil moisture in relation to plant growth. U.S. Dept. Agric. Yearbook of Agric., pp. 358, 1955.

16. Hagan, R.N., Factors affecting soil moisture-plant growth relations. XV Int. Hort. Cong. Wageningen, Netherlands, 1955.

17. Gardner, W.R., Soil Properties and Efficient Water Use, An Overview, Chapter 2 A, In Taylor et al., p. 45, 1983.

18. FAO, Soil Survey Investigations for Irrigation. FAO Soils Bull. No. 42, pp. 188. Cited in FAO Bull. No. 42, FAO, Rome, 1979.

19. Jiery, W.A., W.R. Gardner, and W.H. Gardner, Soil Physics, Chapter 2. 5th ed., John Wiley & Sons, New York, 1991.

20. Hiler, E.A. and T.A. Howell, Irrigation Options to Avoid Critical Stress, C. Scheduling irrigation by water balance techniques using computers, Chapter 11 A, In Taylor et al., p. 479, 1983.

21. Stringhamet, G.E. (Ed.), Surge Flow Irrigation. Final Report of the Western Reg. Res. Project W-163 Res. Bull. 515, Utah Agric. Expt. Sta., Utah State University, Logan, Utah, 1988.

22. Stringhamet, G.E. and J. Keller, Surge flow for automatic irrigation, ASCE Irrigation and Drainage Division Specialty Conf, Albuquerque, NM, 132, 1979.

23. Ayers, R.S. and D.W. Westcot, Water Quality for Irrigation, FAO Irrig. and Drainage Paper No. 29, Rev. 1, 1985.

24. Bernstein, L. and E. Francois, Leaching requirement studies. Sensitivity of alfalfa to salinity of irrigation waters. Soil Sci. Soc. Am. Proc., 37:931, 1973.

25. Busch, C.D. and F. Turner, Sprinkler irrigation with high salt-content water. Trans. Amer. Soc. Agric. Eng. (ASAE), 10:494, 1967.

26. Robinson, F.E., Irrigation rates critical in Imperial Valley alfalfa. California Agric., p. 18, Oct. 1980.

27. Bucks, D.A., F.S. Nakayama, and R.G. Gilbert, Trickle irrigation water quality and preventive maintenance. Agric. Water Management, 2:149, 1979.

28. Nakayama, F.S., Water analysis and treatment techniques to control emitters plugging. Proc. Irrig. Assoc. Conf., Oregon, 1982.

29. Montanes, J.L., Automated Irrigation. p. 1, In FAO,[30] Irrig. & Dra. Paper No. 5, 1971.

30. FAO, Automated Irrigation, FAO Irrig. and Drainage Paper No. 5, FAO, Rome, 1971.

31. Loomis, R.S., Crop manipulation for efficient use of water, An Overview, Chapter 8 A, In Limitations to Efficient Water Use in Crop Production, H.M. Taylor, W.R. Jordan, 1983.

32. Land Master Plan of Egypt, Draft Final Report. Vol. 1. Main Report. Euroconsult - PACER, Ministry of Development, Cairo, 1985.

33. Gardner, W.R., Soil-water relations in arid and semiarid conditions. Review of research (Arid Zone Res. XV UNESCO, NIS G1/111), 1960.

34. Doering, E.J., R.C. Reeve, and K.R. Stocking, Salt accumulation and salt distribution as an indicator for evaporation from fallow soil. Soil Sci., 97:312, 1963.

35. Willis, W.O., Evaporation from layered soils in presence of water table. Soil Sci. Soc. Am. Proc., 24:239, 1960.

36. Balba, A.M. and M.F. Soliman, Salinization of homogeneous and layered soil columns due to upward movement of saline groundwater and its evaporation. Agrochimica 13, 6:542, 1969.

37. Balba, A.M. and M.F. Soliman, Effect of kind and concentration of solute on water evaporation and salt distribution in sand columns. Alex. J. Agric. Res., 26:739, 1978.
38. El Gabaly, M.F. and M.N. Naguib, Effect of depth and salt concentration of groundwater of soil, Int. Sodic Soils Symp. Agrok. es Talaj., 14:369, 1965.
39. Wilcox, L.V., The Quality of Water for Irrigation Use. USDA Tech. Bull. No. 962, 1948.
40. Thorne, J.P. and D.W. Thorne, The irrigation waters of Utah. Utah Agric. Expt. Bull. 346, p. 46, 1951.
41. Rhoades, J.D., Quality of irrigation water. Soil Sci., 113:277, 1972.
42. Ayers, R.S., Interpretation of quality of water for irrigation. Expert Consultation on Prog Salinity and Alkalinity, Rome, FAO, 1975.
43. Bryssine, G., Cited in Durand, S.H., The quality of irrigation water. Sols African, 4:53, 1959.
44. Gardner, W.R. and R.H. Brooks, A descriptive theory of leaching. Soil Sci., 83:295–304, 1957.
45. Brooks, R.H., J.O. Goertzen, and C.A. Bower. Prediction of changes in the composition of the dissolved and exchangeable cations in soils upon irrigation with high-sodium water. Soil Sci. Soc. Am. Proc., 22:122, 1958.
46. Balba, A.M. and H. Bassiuni, Studies of salt movement in soils under leaching process using tracer techniques. Isotope and Radiation Res. Vol., 9:71, 1977.
47. Balba, A.M., A quantitative study of the salination and desalination processes of soil columns. Trans. Sodic Soils Symp. Budapest. Tal. es Agrokem., 14:351, 1965.
48. Eaton, F.M., Significance of carbonates in irrigation waters. Soil Sci., 69:123, 1950.
49. Balba, A.M. and A. Balba, Effect of anion composition of the solution phase on the cation exchange reaction. Int. Symp. Salt-Affected Soils Proc. 361, 1972.
50. Talur, S.R., A.L. Page, and N.T. Coleman, The influence of ionic strength and ion pair formation between alkaline earth metals and sulfate on Na-divalent cation exchange. Soil Sci. Soc. Am. Proc., 32:639, 1968.
51. Singh, S.S. and R.C. Turner, Sulfate ions and cation exchange reactions with clays. Can. J. Soil Sci., 45:271, 1965.
52. Bernstein, L., Salt-affected soils and plants. Paris Symp. on Problems of Arid Zones. UNESCO Pub., 139, 1962.
53. Bernstein, L., Salt tolerance of plants, USDA Agric. Inf. Bull. No. 283, 1964.
54. Bower, C.A., G. Ogata, and J.M. Tucker, Rootzone salt profiles and alfalfa growth as influenced by irrigation water salinity and leaching fraction. Agron. J., 61:783, 1969.
55. Roades, J.D., Drainage for salinity control. In Drainage for Agriculture. Agron. Series No. 17, 1974.
56. Rhoades, J., Leaching requirements for exchangeable Na control. Soil Sci. Soc. Am. Proc., 32:652, 1968.
57. Maas, E.V. and G.J. Hoffman, Crop salt tolerance. Evaluation of existing data. Proc. of Int. Salinity Con. Lubbock, TX, U.S., 187, 1976.
58. Maas, E.V., Salt tolerance of plants, 1984, Cited in Ayers and Westcot.[27]
59. Hoffman, G.T., Guidelines for reclamation of saltaffected soils. Proc. Inter. Amer. Salinity Water Management Technology, Juarez, Mexico, p. 49, 1980.
60. National Academy of Science, and National Academy of Engineering, Water quality criteria. U.S. Env. Protection Agency, Washington D.C., Report No. EPA-R, 373-033, 592 p., 1972, In Ayers and Westcot.[23]
61. Pratt, P.F., Quality criteria for trace elements in irrigation waters. California Agricultural Exp. Sta. 46 p., 1972. In Ayers and Westcot.[23]
62. Evans, R.J., I.G. Mitchel, and B. Salau, Heavy metal accumulation in soils irrigated by sewage and effect in the plant-animal system. Progressive Water Technology, 11(4/5):339, Pergamon Press, 1979. Cited in Ayers and Westcot.[23]

63. WHO, Reuse of effluents: Methods of wastewater treatment and health safeguards. Report of a WHO Meeting of Experts. Tech. Rep. No. 517, WHO, Geneva, 1973.
64. FAO, Integrated Farm Water Management, In Irrig. and Drainage Paper No. 10, FAO, Rome, 1971.

CHAPTER 6

Desertification

INTRODUCTION

The incidents of desert expansion at the expense of productive land have repeatedly taken place in several locations around the globe. Kropotkin[1] who traveled in the wastes of Central Asia concluded that the whole of that wide region is now and has been since the beginning of the historic period in a state of rapid desiccation. This rapid desiccation pushed the inhabitants to migrate and invade Europe in the first centuries of this era.

The concept of progressive desiccation prevailed for a long time. It implies that aridity had increased since the warming of the Pleistocien Era (Goudie[2]). In Africa, the concern established by the concept of progressive desiccation as a reason for desert expansion among the Colonial Anglofrench Governments was such that they formed a commission to investigate the problem. The specialized commission reported that "while there was little evidence for climatic retrogression, there was evidence for adverse human practices which produced dune sand reactivation and water table lowering. Population movement toward the coast was regarded as being a response to economic forces" (Jones[3]).

Doubts that climatic changes are the sole reason of desert expansion had started. Another concept emerged since the 1960s to explain the expansion of desert. It implies that this phenomenon is a result of human activities as well as natural extreme events (Stump[4]). In this period the terms desertification and desertization became common.

The term desertification was first used but not formally defined by Aubreville[5] in 1949 as stated by Goudie.[2] Kovda[6] introduced the term "land aridization" to mean several processes and trends that reduce the effective moisture content over large areas and decrease the biological productivity of the soils and plants of the ecosystem.

A definition of desertification adopted by the Desertification Map of the World (UNEP/FAO/UNESCO/WMO[7]) is: "the intensification or extension of desert conditions. It is a process leading to reduced biological productivity with consequent reduction in plant biomass as in the lands carrying capacity for livestock, in crop yields and human well being." Another definition adopted by the UN Desertification Conference,[8] states that "desertification is the diminution or destruction of the biological potential of the land which leads ultimately to desert-like conditions."

Dregne's[9] definition of desertification emphasized the role of man's activities as follows:

"Desertification is the impoverishment of terrestrial ecosystems under the impact of man." It is the process of deterioration in these ecosystems that can be measured by reduced productivity of desirable plants, undesirable alterations in the biomass and the diversity of the micro and macro-fauna and flora, accelerated soil deterioration and increased hazards for human occupancy. Sabadell et al.[10] admitted the possible importance of climatic controls but gave them a relatively inferior role: "Desertification is the sustained decline and/or destruction of the biological productivity of arid and semiarid lands caused by man-made stresses sometimes in conjunction with natural extreme events. Such stresses if continued or unchecked over a long period lead to ecological degradation and ultimately to desert-like conditions."

Le Houerou[11] defines desertization as the internal irreversible growth of desert areas in the arid regions which has not been shown in the near past. These symptoms are characterized by the diminution of the permanent vegetative cover and its concentration along the groundwater sources with the formation of sand dunes, desert platforms, or gravelly plains. Permanent vegetation covers only about 5% of the area and mostly does not exist.

Le Houerou[12] relates the growth in the desert area in the arid regions to dryness. He defines dryness as diminution in precipitation relative to its recorded average during several years which causes a considerable decrease in the agricultural production extending among vast areas. One of the well-known drought waves is that of the Sahel region which lasted about 15 yr (1970–1985). However, three drought incidents had taken place in the same region in the past period 1900 to 1950 (1895–1905, 1910–1916, 1938–1943) (Bernus and Savonnet[13]). These desiccation cycles are known at present with their close relation with the ENSO, the variations which take place in the south of the Pacific Ocean. Soils and plants in the arid regions have been accustomed along thousands of years with these conditions. Le Houerou[12] concluded that the increase in the population is the main cause of desertization. Desiccation cannot be the sole cause. It is an auxiliary factor that made the circumstances more likely. He tried to prove this concept by showing that where the population increase was enormous, desertization was most active.

UNCOD Secretariat[14] discussed the causes of desertification and pointed out the following:

- The shifting limits of dry land climates indicating that in the Sahel region 1000 years ago, semiarid or subhumid conditions had encroached northward into what is now arid.
- One cannot accept with confidence that the climate change is the cause of desertification.
- It may be possible that man's activities have influenced the climate accentuating dryness, but these activities are not likely to be prime causes of any general deterioration of dry land climates.
- Climatic fluctuations contribute to desertification.
- The dry lands ecosystem is sensitive and fragile. The soils are poor in organic matter and nutrients. Thus, any use of these lands without taking account of their limitations constitutes misuse and leads to desertification.
- Desertification is a product of the interaction between man and a difficult environment.

Rozanov[15] considers that the UNCOD definition of desertification is not adequate from the technical point of view:

- It is not clear what "desert-like conditions" are, as there are some deserts without any plant life, while others are rather green.
- Any degradation of a territory is understood as desertification under UNCOD's definition.
- There is no distinction between desertification and periodic droughts.

- There are no clear-cut measurable and objective criteria of desertification. Thus, the adopted definition does not provide for concrete and precise parameters for quantitative assessment, monitoring, and control of the process per se.

Rozanov[15] suggested the following definition for desertification:

"Desertification is a process of irreversible change of soil and vegetation of dry land in the direction of aridization and diminution of biological productivity, which in extreme cases, may lead to total desertification of biospheric potential and conversion of the land into desert."

According to Rozanov,[15] his definition requires precise understanding of its terms as follows:

- "Irreversible change" is such a change of soil or vegetation that requires either man's ameliorative interference or very long (decades or even centuries) natural processes for restoration.
- "Dry land" is a territory in tropical, subtropical or warm arid, semiarid, or seasonal subhumid climate.
- "Desert" is a dry land which is almost totally devoid of vegetation and developed soil.
- "Biological productivity" is the annual production of biomass expressed in tons per hectare per year.
- "Aridization of soil" is a change of soil toward decreased ability to supply plants with available water.
- "Aridization of vegetation" is an increase of xerophilous species in the vegetation, together with general diminution of vegetation density and biological productivity.

This definition includes both natural processes of desertification and those induced by human action.

SOIL DEGRADATION

The net result of desertification is the change of the productive soil to unproductive, which has been known among agronomists, especially soil scientists, for some time by the term deterioration or degradation. Thus, soil degradation has been tackled by the UN Food and Agriculture Organization[16] (FAO) for a long period of time. The terms degradation or deterioration signify the reduction of the land productivity inherent to changes in the soil physical, chemical, or biological properties. FAO concentrated its activity—in this field—on evaluating the degree of deterioration, analyzing the processes that caused it, and measuring the rate of deterioration caused by each process. Reclamation of deteriorated soils received considerable attention.

Soil degradation as defined by FAO/UNEP/UNESCO[16] is "a process which lowers the current and/or the potential capability of soil to produce (quantitatively and/or qualitatively) goods or services. Soil degradation is not necessarily continuous. It may take place over a relatively short period between two states of ecological equilibrium."

The processes of soil degradation are mainly water erosion, wind erosion, salinization and/or sodication, chemical degradation, physical degradation, and biological degradation.

Units for Measuring Soil Degradation

- Soil erosion by water: soil loss in t/ha/yr or in mm/yr.
- Soil erosion by wind: soil loss in t/ha/yr or in mm/yr. (t/ha/yr = 0.66 mm/yr and 1 mm/yr = 15.0 t/ha/yr assuming soil bulk density 1.5 g/cm³).
- Salinization: increase of electrical conductivity of saturated paste extract at 25°C, in dS m⁻¹.
- Sodication: increase of exchangeable sodium percentage

$$(ESP) = \frac{Exch.\ Na}{Cation\ exch.\ capacity} \times 100, \text{ in percent/yr}$$

Both salinization and sodication refer to the soil layer 60 cm depth.
- Chemical degradation:

  - acidification: decrease of base saturation in

  $$\text{percent/yr (base saturation} = \frac{Total\ exch.\ bases}{Cation\ exch.\ capacity} \times 100)$$

  - increase in toxic elements mg kg⁻¹. As both chemical degradation processes are often active mainly in the topsoil, the soil layer 30 cm depth is referred to, in order not to dilute the effect by taking another 30 cm of possibly little-affected soil into account.

- Physical degradation: increase in bulk density or in permeability in cm/h/yr. For physical degradation the 60 cm soil layer is referred to.
- Biological degradation: decrease in humus in percent decrease/yr. As biological degradation is also very much a topsoil phenomenon, the 30 cm depth is referred to.

In order to compare the processes of soil degradation together, a broad classification of the seriousness of each process was made in such a way that the classes are approximately equal.

The following classification is an example (FAO/UNEP/UNESCO[16]):

Degradation Classes

Water Erosion (E)

	Soil Loss		
	t/ha/yr	or	mm/yr
None to slight	10		0.6
Moderate	10–50		0.6–3.3
High	50–200		3.3–13.3
Very high	200		13.3

Wind Erosion (W)

Class limits for soil loss are as for water erosion.

	Salinization (Sz) Increase in ECe 60 cm layer	Sodication (Sa) Increase in ESP 60 cm layer
None to slight	2 dS m^{-1} yr^{-1}	1 ESP/yr
Moderate	2–3 dS m^{-1} yr^{-1}	1–2 ESP/yr
High	3–5 dS m^{-1} yr^{-1}	2–3 ESP/yr
Very high	5 dS m^{-1} yr^{-1}	3 ESP/yr

Chemical Degradation (C)

i) If the base saturation is less than 50%.

	Decrease of base saturation
None to slight	< 1.25%/yr
Moderate	1.25–2.5%/yr
High	2.5–5.0%/yr
Very high	< 50%/yr

ii) If the base saturation is more than 50%.

	Decrease of base saturation
None to slight	< 2.25%/yr
Moderate	2.25–5.0%/yr
High	5.0–10%/yr
Very high	> 10%/yr

Physical Degradation (P)

i) Increase in bulk density (% change) with reference to initial level.

	Initial level			
	1.0 g/cm^3	1.0–1.25 g/cm^3	1.25–1.4 g/cm^3	1.4–1.6 g/cm^3
None to slight	< 5	< 2.5	< 1.5	< 1
Moderate	5–10	2.5–5.0	1.5–2.5	1–2
High	10–15	5.0–7.5	2.5–5.0	2–3
Very high	> 15	> 7.5	> 5	> 3

ii) Decrease in permeability (% change) with reference to initial level (cm/hr).

	Initial level		
	Rapid (20 cm/h)	Moderate (5–10 cm/h)	Slow (5 cm/h)
	% change/yr		
None to slight	2.5	1.25	1.0
Moderate	2.5–10	1.25–5	1–2
High	10.0–50	5.00–20	2–10
Very high	> 50	> 20	> 10

Biological Degradation (B)

	Decrease in humus (0–30 cm layer)
None to slight	< 1/yr
Moderate	1.0–2.5%/yr
High	2.5–5.0%/yr
Very high	> 5%/yr

DEGRADATION* VS. DESERTIFICATION

Though both terms "degradation" and "desertification" end up with partially or totally unproductive soil, the two terms are not synonyms. Degradation used to be concerned with changes in the soil physical, chemical, and biological properties which affect the soil as a medium for plant growth (FAO[16]). Desertification pays more attention to environmental factors which affect the soil productivity. Accordingly, desertification may take place due to sand dune movement or deforestation. Degradation processes as stated above do not include, at least directly, both processes. Both terms give an important role to man's activities.

In Rozanov's[15] opinion, desertification is a result of (1) natural processes as a change of climate in the direction of aridity, or of certain processes of "normal" geological development of a territory in the condition of unchanged arid climate, and (2) desertification also may be induced by man's action connected with a disturbance of the natural equilibria of arid ecosystems as a result of intensified use of the land beyond its potentiality.

He concluded that indicators of soil degradation—water and wind erosion, salinization, sodication, waterlogging under irrigation, destruction of structure, depletion of humus and nutrients—although they accompany desertification in various situations in different combinations, are not specific to it. Thus, they cannot be used for its diagnosis or assessment. He further stated that the decrease of biological productivity, if it is not connected

---

*   Degradation used to describe the result of leaching that leads to soil acidification.

with soil aridization but is a result of some temporarily acting causes, e.g., drought, it cannot be considered diagnostic.

To quantitatively determine the desertification process, the two inter-connected parameters,

a) degree of aridization of vegetation and
b) degree of aridization of soil, should be taken as the basis for the determination.

Two different methods may be used:

a) Comparison of the state of the same area at different times.
b) Comparison of the state of two different areas at the same time.

Desertification has qualitative and quantitative manifestations in the natural state of the landscape different from in the cultivated lands. Thus, assessment of desertification should be made separately for the natural ecosystems and for artificially made agro-ecosystems.

Accordingly, Rozanov[15] excludes the irrigated lands from the assessment as they are—in his opinion—not subject to desertification (unless threatened by shifting sands coming from outside), even if there is soil degradation (such as erosion, salinization, sodication, etc.). Such degradation cannot be taken as an indicator of desertification.

Degree of Desertification

It is common to express the degrees of the natural processes by grades of intensity: slight, moderate, strong, and very strong. These grades were used in classifying the intensity of desertification as stated above. According to Rozanov,[15] the content of each of these classes cannot be the same in all natural zones of the globe, e.g., there cannot be slight or moderate desertification in zones geographically denoted as deserts. Also, slight desertification will have different quantitative manifestations in tropical xerophytic forests and savannas.

The degree of aridization of vegetation in natural ecosystems, including natural pastures, may be made by the quantitative determination of:

a) ratio of climax and invading species in plant cover composition,
b) ratio of xerophilous and mesophilous species in plant cover composition,
c) density of the plant cover,
d) biological productivity.

Determination of these parameters requires special botanical investigations.

Biological productivity, the annual increment of the above ground phytomass, may be taken as an approximate assessment for desertification. In arable lands, the annual yield which is proportional to the total biomass increment, may be taken as an indicator of biological productivity.

Appropriate scales of degree of aridization or desertification (current state of desertification) for different physico-geographic conditions and natural zones can be constructed on the basis of detailed studies of biological productivity and other parameters of the state of vegetation. For each natural zone, the quantitative parameters of the state of vegetation and its degree of aridization will be different in accordance with the differences in natural and cultural plant associations.

After discussing the degree of soil aridization and its relationship with certain physical properties and water regime of soils, Rozanov[15] concluded that:

a) The annual supply of available water may be a convenient indicator of soil state and degree of aridization.

b) The annual deficit of soil water may be taken as an indicator of the degree of soil aridization.

c) The water regime of a soil taken in certain hydrological terms may be used as a criterion of the degree of soil aridization.

Because avoiding desertification is easier and cheaper than curing desertification areas, assessment of desertification risk including the degree of risk are needed. Rozanov[15] defined the concept of desertification risk as follows:

"The degree of potential development of the desertification process exceeding the permissible limits of ecological equilibria changes in a territory during its use by man."

He added that natural processes of desertification are excluded from this concept of risk.

The degree of risk varies with:

- Kind of land use.
- Soil characteristics.

Grades of desertification risk may be as follows:

- Very slight risk: desertification under a given land use is not excluded but its probability is negligible.
- Slight risk: desertification is possible but its degree and extent will not be substantial.
- Moderate risk: desertification is probable and its degree and extent will be substantial.
- Strong risk: desertification is unavoidable and its degree and extent will be substantial.
- Very strong risk: desertification is unavoidable and its degree and extent will be catastrophic.

Desertification takes place gradually and under variable conditions. Its processes may also vary but the result is always the same: the change of the productive land to unproductive and the expansion of the desert area.

Combating Desertification, A General View (UNEP[17,18,19])

The UN Conference on Desertification[8] produced the Plan of Action which was endorsed by the UN General Assembly. The aim of the Plan of Action is (1) to stop desertification expansions, (2) to reclaim the desertified areas where possible, and (3) to develop national capabilities to combat desertification.

The Plan of Action emphasized (1) the application of existing knowledge in science and technology to develop programs aiming at avoiding further desertification, (2) reclamation desertified lands, and (3) properly using and developing the resources of areas prone to desertification.

Tolba[20] (Dr. M. Tolba, past executive director of UNEP) stated that a comprehensive national plan to combat desertification should comprise two basic elements:

a) monitoring and assessment, and
b) land use policy and planning.

The implementation of such national plans requires:

*   Training and education.
*   Legislation concerning water resources and rights, systems of land tenure.
*   Efficient machinery for implementing the laws.
*   Attention to socio-economic aspects to ensure support for programs designed to preserve the environment.
*   Application of scientific and technological knowledge.

In pasture land, deterioration starts with excessive grazing in drought periods. The consumption and deterioration of the palatable perennial plants are accompanied with the increase of unpalatable plant growth. With the decrease and death of plants in the drought periods, the bare areas expand followed by the deterioration of the soil surface. The rain runs off as it does not penetrate through the soil, thus eroding the soil surface. The soil fertile surface layer is lost causing further decrease of soil productivity. The soil erosion is especially important in sloping areas. Over time, erosion extends and the soils which were productive pasture lands lose their productivity and are added to the desert (UNCOD Secretariat[14]; Bennet[21]).

In the rain-fed cropland, desertification occurs in barren areas either after crop harvesting or before sowing. As long as the surface is bare, erosion takes place.

The irrigated land is subject to salinization or sodication as a result of the lack of an efficient drainage system.

Several other natural phenomena such as sand dunes creep or deforestation are major causes of desertification.

DESERTIFICATION DUE TO SAND MOVEMENT

In arid and semiarid regions, the action of wind upon unconsolidated surficial materials can give rise to several problems.

Wind erosion can result from mismanagement of cropland, overgrazing of pasture land, and excessive cutting of trees and bush cover. The coarser particles of soil carried by the eroding wind usually move close to the ground surface until their movement is arrested by physical obstructions, thus large heaps form. These large heaps are described as sand dunes. When large dunes are formed, they create problems by creeping over farm land, roads, railways, and even whole villages. Coastal dunes may be blown inland, especially if protective plant cover is destroyed.

The total area of land vulnerable to desertification due to sand movement was evaluated from the Desertification Map of the World[7] by Balba[22] and found to be about 5,770,000 km^2. This area was divided among Africa 3,190,000 km^2, Asia 60,000 km^2, Australia 1,370,000 km^2, and South America 460,000 km^2.

The sand dune is a hill of "sand." Its height differs from a few meters to tens of meters. It is usually formed of spherical particles having a diameter of 0.05 to 3.0 mm.

Two main kinds of sand dunes are known: Coastal and Continental dunes (UNCOD Secretariat[14]).

The coastal dunes are usually formed from coarse sand particles whitish in color due to its carbonates, chlorides, and sulfates content. They are usually poor in clay and in nutrient elements. They extend lengthwise along the coasts of oceans and seas. They usually accumulate on the coast as a result of the tide.

The continental dunes are usually formed of quartz particles, reddish in color, mixed with feldspars and carbonates.

The forms of the sand dunes differ according to their origin and their accumulation conditions. They may assume several forms having several terms to describe them. Mader and Yardley[23] classified the dunes into primary, secondary, and complex dunes.

The primary dunes may be simple, migrating, or transverse forms. The secondary may be longitudinal forms produced by morphological modification and migration. The complex or tertiary dunes form star dunes produced by merging of transverse and longitudinal elements.

For simplicity, the sand dunes are classified into (Imbaby and Ashour[24]; Schou[25]):

i.   Longitudinal dunes: long dunes formed from several parallel lines. They are oriented parallel to the wind direction and extend excessive distances. This form is familiar in the Great Desert of Africa (the Sahara).

ii. Barchan dunes: or Transverse dunes are perpendicular to the wind direction and have the crescent or elliptical form.

iii. The Sandy Shades dunes: are formed when winds loaded with sand are faced with an obstacle such as a rock, thus the wind is deflated into two directions at both sides of the rock. A cyclonic wind is formed at each side of the obstacle causing a slower wind speed. Sand particles are deposited because of the slower wind speed forming two heaps of sand which soon join together into one heap with the deposition process of sand.

iv. Coastal sand dunes: when the wind blows from the sea side it moves the coastal sand and deposits it into dunes parallel to the coast. Plants growing along the coast prevent the sand movement.

In the Desertification Map of the World,[7] the area of a very high degree of desertification hazards due to sand movement (the regions subject to very rapid desertification if existing conditions do not change) was about 44.0 M ha, out of which about 43.5 M ha are in Africa and 50,000 ha in Asia. As afforestation of these desertified areas requires an adequate level of moisture in the soil, success in semiarid regions to combat sand dunes is more likely than in arid ecosystems. The area with a very high degree of desertification hazards due to sand movement in the arid regions in the above mentioned map amounts to about 15 M ha. Thus, it may be more logical to exclude this area from a global program to combat desertification due to sand movement. The remaining area amounts to 29 M ha, mainly in Africa (Balba[22]).

It should be pointed out that controlling wind-blown sand does not affect the sand dune movement though practices of both cases may be similar.

Measures to Combat Desertification Due to Sand Movement

Preparatory (UNCOD Secretaria[12])

The first step is to survey the area. Several means might be used, namely remote sensing, aerial photography, and ground marking. The obtained maps should show the distinct topographic features of the location, the vegetative cover, water resources, and other characteristic necessary details. Climatic records from the nearest meteorological station should be secured. Traditional social values and habits of the inhabitants should be taken into consideration and their cooperation in favor of the project is necessary.

The evidence obtained from the satellite images, aerial photographs, maps, reconnaissance, and semi-detailed surveys of the area should show the types of desertification and its degree of severity in each part of the area and the relative vulnerability of the ground units to further desertification and economic activities in the area assessed. Finally the plan of action to counteract the causes of the problem and to reestablish the disturbed population can be formulated.

Sand movement may be of two types: (1) blowing sand caused by winds, and (2) desert sand dunes creep.

There have been several approaches to controlling blowing sand (Watson[26]; Fryreer[27]; Chepil and Woodruff[28]) in areas where sand moves by wind.

Control of Blowing Sand Caused by Winds

Deposition of the Sand

Reducing the wind load of sand by decreasing the wind transporting capacity can be achieved by several means.

Excavating Ditches and Trenches (Watson[26])

For the ditch to be fully effective, it must be wider than the maximum horizontal jump of saltating sand grains at about 3.0–4.0 m wide.

Its depth should be enough to prevent transportation of sand by wind from the bottom of the trench.

Clearing the deposited sand from the trench may be necessary. The other alternative is to excavate new ditches. The excavated material should be removed to a location where it will not present a hazard.

Windbreaks

The windbreaks whether, they are fences or rows of trees, create a barrier to the wind and produce areas of reduced sand carrying capacity ahead of and behind that barrier (Balba,[22] Watson[26]).

Fences

The fences may be solid or porous. Deposition of sand takes place in a zone at a distance from 0.4–2.0 times the high of the solid barrier. In the case of porous barriers deposition of sand occurs in the lee, 4 times as wide as the fence height. Percent voids to total area of the porous fence as well as the arrangement of the spaces affect the efficiency of the fence. The efficiency of the barriers in controlling the sand deposition is affected by their spacing and alignment relative to the wind direction.

Vegetative Belts (Balba,[22] Jensen[29])

They are rows of trees grown across the direction of the prevailing winds to arrest their force and to prevent the adverse effects resulting from such force.

In order to obtain upward deflection of the wind movement (as downward deflection will damage the soil), the windward side of the windbreak should have a sloping side inclined upward toward the top of the windbreak. Also, studies have shown that spacing the shelterbelts can be 20 times the height of the tallest growing trees. The spacing of rows in a 5-row windbreak depends on the growth habits and size of the plants forming it. Generally, 3–4 m will be satisfactory between the middle and its adjacent inner rows. One row windbreak effectiveness depends on environmental conditions. Studies showed that narrow belts of 1–3 rows have an effect of considerable efficiency. In arid conditions windbreaks 2–5 rows are economic and efficient. An intensive system of windbreaks should be set up in a chessboard pattern, oriented according to the direction of the prevailing winds. The main shelterbelts are recommended to be located at 200–300 m distances, while the secondary shelterbelts, perpendicular to the first, can be located at 500–600 m distances.

Selection of trees for windbreaks depends on their characteristics. They should withstand drought, salinity, low soil fertility, and fluctuations in surface temperature. They must have deep root systems capable of growing vertically to reach the moist layers in the sand and/or have considerable horizontal ramifying spread that allows for efficient use of surface rain and dew. They should be evergreen, fast growing, and grow to considerable heights. Their timber should withstand the blowing winds.

However, the selection of the appropriate species is critical in order to minimize the risk of infestation by disease or pests. Growth of trees in arid and semiarid environments is usually weak. Long-term effectiveness of nonirrigated tree belts is questionable.

In warm semiarid and arid environments, Tamarix and Eucalyptus species have proved successful. Also, *Casuarina* spp., *Cupressus* spp., and *Melaluca orififolia* are commonly used for windbreaks (Watson[26]).

Comparison between vegetation plots and fences showed that fences are more efficient at trapping sand. This is partially due to the initial period required before the plants become established. Once the grass plots are established they are much easier to maintain than trees.

A difficulty encountered in sand stabilization by plants is to find plant species that can withstand severe drought. Also, high rates of sand accumulation or deflation, excessive grazing, high soil salinity, and lack of irrigation water are problems encountered in the biological control of sand movement. In general, it is recommended to depend on indigenous plant species.

Enhancing Transportation of Sand

Enhancing the transportation of sand is another means to decrease the sand load of wind before reaching the area to be protected. Watson[26] pointed out several means for this purpose:

- Shaping the land surface to increase the wind velocity over the soil surface.
- Mantling the land surface to enhance surface creep and sand grain saltation.

Reduction of the Sand Supply (Watson[26])

In cases where enhancing the transportation of sand is not effective, attempts were made to decrease the source of sand that could be transported by winds. This may be achieved by several means: 1) spreading small gravel (about 4.0 mm) on the sand area, thus wind is not able to transport the sand, and 2) stabilizing the sand with vegetation, chemicals or asphalt as will be discussed in the following pages.

Combating Sand Dune Movements

As stated above, success in arresting sand dunes in arid ecosystems especially by afforestation is not likely to succeed. However, there are several procedures to combat sand dunes movement. The effectiveness of a procedure depends on the environmental conditions.

With the increase in man's ability in earthmoving using various powerful mechanical means, the idea of moving the sand dune came up to eliminate the problem. However, to where the excavated sand would be transported is a physical problem that should be solved. Evidently expenses are enormous because of the tremendous amounts of sand that would be excavated and transported. Partial removal of a dune is hazardous as it may increase the mobility of the rest of the dune.

Mechanical manipulation of the dune by reshaping it, trenching, or surface treatment, may be useful provided they are undertaken by specialists (Watson[26]).

Sand Dune Stabilization

The most common procedure to control sand dunes movement is to immobilize them. Immobilization is attempted by mechanical, chemical, and biological procedures.

Mechanical Fixation

In regions where fluctuating wind directions are prevailing, dry materials are used for fencing in a checkerboard system. When the prevailing wind has essentially a fixed direction, parallel lines of the dry materials are established. The surface of the dune might be mulched with dry grasses and the like. Evidently this procedure is limited to small areas.

Chemical Fixation

Several chemicals are used to stabilize the sand dunes. Among these materials are crude oil (asphalt) and synthetic rubber (latex) elastomeric polymer emulsion, polyvinyl alcohol solution, polyvinyl acetate copolymer, and sodium acetate (Bhimaya[30]). Chemical immobilization is recommended in cases where labor cost is high and chemicals are available, as in oil countries and in large areas that need to be fixed in a short time. However, fixation with asphalt is weak in the case of coarse sand and saline conditions. A modification of the procedure is to stabilize strips of the dune instead of the whole surface of the dune. The procedure is mechanized and carried out in Lybia[31] and Saudi Arabia (Kerr and Nigra[32]).

Lyles and Schrandt[33] suggested the formation of a surface salt crust by spraying saline water over the sand surface. Crystallization of the salts upon water evaporation binds the sand grains. However, cracks in the crust may initiate rapid scouring of the underlying sand. Also, runoff water may dissolve the salt crust and hot dry conditions may disintegrate it.

Biological Fixation (Afforestation)

It is common practice—in this procedure—to start with planting grasses, followed by bushes and then by trees. However, in several areas afforestation starts after the mechanical fixation. Plant species used for afforestation must be able to withstand drought, salinity, low soil fertility, and fluctuations in surface temperature. Other characteristics stated above for plants used for establishing fences are also required in plants used for biological fixation of sand dunes.

Attention is drawn to the fact that in order to provide the necessary plants, nurseries are generally necessary.

In general the following points require full consideration in controlling sand dunes (Balba[22]; Balba[34]):

- The whole area should be considered and not the eroding dune only.
- Modification of the surface of the eroding dune can be considered. This is carried out by mechanical reshaping of the dune to remove projections which cause wind turbulence.
- Afforestation should start after the rainy season, when sand is wet.
- Irrigation of tree seedlings in dry seasons is important. Ten liters of water per tree seedling every 7, 17, or 30 days, according to the availability of water are recommended. If water is not available in sufficient quantities, it is advisable to provide irrigation at least during the first 2–3 months.
- Fertilization of plants is recommended.
- Spacing of trees is about 2.5 – 3.0 – 3.5 m.
- Biological fixation using grasses and shrubs also is practiced without preceding mechanical or chemical fixation.
- Because of the expensive cost of mechanical and chemical fixation it is often advisable to depend on afforestation. The length of stock trees should be increased to 5 m in this case.
- Chemical fixation is recommended when the cost of labor is high, chemicals are available, and when large areas need to be stabilized in short periods.
- When a specified area of value is threatened, work is initiated close to the area to be protected with the work gradually progressing toward the sand source area. On the other hand, if there is a well defined sand source area, such as a beach, the protection work is often started close to the source. A combination of the two approaches may be incorporated.

It is necessary to point out that combating desertified areas or immunizing areas from desertification due to sand dune encroachment requires positive cooperation of the inhabitants of the area. Also, legal measures should be applied and enforced.

According to Finkel[35]:

- The control of wind-blown sand may have no effect on sand dune movement.
- The source area of blowing sand may be very distant. Accordingly, stabilization of the sand source may prove impractical or impossible.
- Ditches are useless as a means to check sand dunes since they can be filled with sand.
- Mechanical disturbance and physical barriers to sand movement are ineffective.
- Vegetative stabilization techniques are usually impractical because they require irrigation.
- Techniques that reduce the height or volume of the dunes may merely aggravate the problem by increasing the rate of advance.

DESERTIFICATION OF RANGELAND

Where rainfall is scanty and highly variable, arid grasslands prevail. These lands are suitable for a certain degree of grazing. All features of life and human activities are controlled by seasonal climatic variations and the availability of water sources to satisfy the needs of man and his herds. The inhabitants of these areas live under the consistent threat of rainfall failure.

Rangeland Vulnerability to Desertification (UNDESCON Secretariat[12]; UNEP[17,18])

Vulnerability to desertification and the severity of its impact are governed by:

- Climate: the lower and more uncertain the rainfall, the greater the potential for desertification. Seasonality of rainfall also is of importance.
- Soil texture and structure and type of vegetation.
- Liability of desertification is a function of pressure on land use as reflected in density of population or livestock.

Characteristics of Rangeland in Dry Ecosystems

- Biological productivity matches moisture condition; both are highly variable.
- Spottiness of rainfall and runoff characteristics result in patches of plants.
- Indigenous plants and animals have acquired efficient mechanisms for resilience. They must not be damaged.
- Recovery in dry ecosystems is slow because of the low diversity of plant species and the rarity of seasons with sufficient moisture for recovery.

Rosenzweig[36] cited by Coe[37] described the relationship between net above ground annual primary production (NAAP) and the annual actual evapotranspiration (AE) as follows:

$$NAAP = \log_{10} AE \ (1.66\pm0.27) - (1.66\pm0.07)$$

Effect of Heavy Grazing

Direct Effects

- Heavy grazing removes especially the leaves of grasses, thus curtailing photosynthesis and seeding. The most palatable grasses are the first to be damaged, thus their frequency in the community is reduced.
- The devastation around the wells often attributed to overgrazing, but is actually the result more of trampling than of overgrazing. Thus the hooves pound, powder, or puddle the bare ground. When rains come, the soil is puddled and becomes impermeable.

The nitrogen from droppings of the herds is denitrified and might contaminate the water.

Indirect Effects

- Unpalatable plant species resistant to damage factors expand their growth cover. Poisonous plants may become dangerously common. Woody shrubs may invade the pasture. A general principle is that the most valuable plant communities are often the first to suffer.
- The grass-eating termites seriously compete with sheep for grasses in droughts.

Measures to Combat Desertification of Rangeland (FAO[38]; Balba[22]; UNEP[17,18])

1. Surveys are essential to determine the variety of dryland pasture under differing seasonal conditions, the requirement of pasture for successful regeneration under grazing, and the dimensions of the grazing impact of the proposed system composed of certain animals in certain numbers.
2. Means to assess the land carrying capacity.
3. Intensive reclamation, as a planting program, is feasible in restricted areas where the physical processes of desertification threaten valuable cropland.
4. Conservation measures should be introduced for control of grazing access to dryland ranges.
5. The dual role of perennials as surface protectors and as fodder during drought should be taken into account.
6. The consumption of firewood should be controlled. Reforestation programs must be encouraged as well as introducing other sources of energy.
7. Grazing strategies should incorporate possibilities for deferred or rational grazing and for establishing protected reserves such as seed reservoirs, grazing reserves in the event of drought, and plant and wildlife refuges in which genetic varieties can be conserved. These parts of the rangelands might be fenced to limit stock movement through them.
8. Simple water-harvesting schemes such as trenches and flood dikes should be developed.
9. Grazing strategies should give attention to localized concentrations of livestock such as on long tracts, around salt licks, watering points, and settlements and measures should be taken to avoid intensive local grazing and trampling.
10. Optimum carrying capacity should be based on a maximum consumption corresponding to one third of the maximum herbaceous biomass.

11. When the grazing should be reduced, several steps should be taken:

    a) Improve transport facilities.
    b) Establish marketing outlets.
    c) Improve productivity of the herd through breeding.

12. The pastoral communities need help such as insurance of their herds, subsidies, and relief in drought periods.
13. Where cropland adjoins rangeland, mutual support of the farmers and the rangers is beneficial for both groups.
14. Improve marketing facilities.
15. Accounts must be taken of traditional social values.

It is necessary to gain willing participation of the local communities. It might be necessary to create incentives and establish pilot areas.

Techniques that increase range productivity should be followed:

- Disease control and animal health improvement.
- Pasture regeneration through grass seeding and forage plantation when the climatic conditions allow.
- Genetic improvement to obtain drought-resistant varieties of grasses.
- Fodder and supplemental production.
- More adequate water distribution.

Total Catchment Management

According to Cunningham,[39] the resources of an area should be managed as a whole. He stated that this "total catchment management approach" involves the coordinated use and management of land, water, vegetation, and other physical resources and activities within a catchment to ensure minimal degradation and erosion of soil and minimal impact on water yield and quality and on other features of the environment.

Specifically, total management aims to:

- Encourage effective coordination of policies and activities of relevant departments, authorities, companies, and individuals which impinge on the conservation, sustainable use, and management of the states' catchments including soil, water, and vegetation.
- Ensure the continuing stability and productivity of the soils, a satisfactory yield of water of high quality, and the maintenance of an appropriate protective and productive vegetative cover.
- Ensure that land within the country's catchments is used in its capability, in a manner which retains as far as possible options for future use.

Areas Affected by Rangeland Desertification

Desertification of rangeland especially in arid and semiarid ecosystems is wide-spread.

a) The UN Conference on Desertification[8] and the FAO[38] approved Dregne's[40] estimate that range deterioration is as much as 3.2 billion ha, most of which, 2 billion ha, is in Africa.

b) Considering that the area of the semiarid region is about 6,715,759 square miles or about 1 billion ha as estimated in Arid Land News Letter,[41] it is clear that deteriorated semiarid rangeland area is much less than Dregne's estimate.

DESERTIFICATION DUE TO IRRIGATION AND WATERLOGGING*

In arid and semiarid regions, irrigation is the backbone of intensive agriculture. Man has always been trying to discover water resources and means to make use of them. Wells have been dug, dams have been constructed, and canals have been extended to deliver the water to where it is more efficiently utilized. However, these efforts were not always successful. Instead of obtaining more food, introducing irrigation in several areas resulted in soil deterioration. Several projects have been constructed to store and distribute irrigation water without consideration of drainage of the irrigated land. Waterlogging thus takes place from canal seepage, overirrigation, and poor internal drainage. As the water table rises, salts accumulate at the soil surface and as soil water evaporates it leaves its load of salts behind.

Salt-affected soils cover extensive areas in many countries in arid and semiarid ecosystems of all continents causing considerable problems regarding the natural environment and national economics. The expectations from irrigation were very high, but short-sighted projects have neglected the effects on soils and plants of inadequate drainage, although technical experience in a number of countries has been to avoid such miseries. It has been frequently stated that the irrigated land lost annually to desertification is about equal to the area of land newly brought under irrigation each year. Large investments are involved in such projects.

Soil salinization, sodication, and waterlogging are complex phenomena related to water movement in the soil profile and affected by several factors.

Magnitude of the Problem**

In considering the problem of desertification due to salinization, sodication, and waterlogging and formulating remedial programs, comprehensive appraisal of the economic factors involved should be carried out. They are based on estimation of the areas suffering from and subject to desertification due to salinization, sodication, and/or waterlogging.

Another way to look at the problem is to consider the present and projected irrigated areas in the arid and semiarid regions as subject to salinization, sodication, or waterlogging unless they are equipped with drainage to avoid these hazards (FAO[38]).

Measures to Combat the Problem

To avoid problems in irrigated areas, integration of research and planning is called for. Planning should be based on extensive studies of groundwater, soil, irrigation system

---

* More about the salt-affected soils is presented in Chapter 2 of this book.

** Estimates of salt-affected soils in the world are discussed in Chapter 2 of this book.

suggested for the project, quality of water to be used for irrigation, cropping pattern, and climatic conditions (FAO[38]; Balba[42]).

From these studies, the components of the water and salt balance in the project area are known. Thus, it becomes possible to know the rate of groundwater table rise toward the soil surface, and the time after which the groundwater reaches the level which causes waterlogging. Also, the salt balance can be calculated from which the net amount of accumulated salts would be known.

Such investigations should show not only the need for a drainage system, but also how efficient the whole planned scheme including the planned drainage system is to overcome the expected problems of soil salinization and waterlogging (FAO[38]; Balba[42]).

COMBATING DESERTIFICATION IN RAIN-FED CROPLAND

Rain-fed cropping systems comprise several types as determined by climate and other environmental conditions. Each is marked by its characteristic crops, technology, and cultural setting. Each is vulnerable to desertification, which takes on distinctive forms in each setting and calls for distinctive measures to combat it.

In the Mediterranean, forests have given way to dryland shrubs or to bare earth with soils sometimes stripped completely, to uncovered, calcareous crusts or naked rock. In many places runoff has become ephemeral. Subsidence of groundwater tables, depletion of aquifers, and decline in water quality are frequently encountered (UNEP[19]; Balba[34,43]). In the sub-tropical to warm-temperate regions, the plains that are characterized by absence of trees are subject to degradation of soil through wind and water erosion. In cool temperate semiarid regions, which largely consist of open plains, erosion is the characteristic form of desertification; structureless and light-textured soils, often lying on carbonate or hardpan layers, are most affected during dry winter or late summer. In tropical semiarid summer monsoon regions, typified by the Sudanian belt, open savanna woodland is cleared, usually by burning, to provide seedbed, although clearance is not complete and many trees may be left standing. The pattern is generally that of shifting agriculture. Four to five years of continuous cropping are followed by abandonment, when successional regrowth of acacia may be harvested, or grazed by cattle, with the growth of grass encouraged by burning. Desertification in these systems often appears as a marked decline in fertility and a deterioration in the soil structure. Rainfall is often intense, causing pluvial erosion of cultivated surfaces. Soil surfaces become puddled and soil structure severely damaged. The dry spells following the rainfall bake a crust on the surface, hindering germination and development of seedlings. During the dry winters, wind erosion lifts clouds of dust from these lands (UNEP[18]; Adu[44]).

Soil Conservation Measures to Combat Desertification (UNEP[17])

A first step in formulating a plan is to map land types and land use at a scale appropriate to cropping. The land units mapped should be classified according to potential use as determined by soil and water resources and by topographic hazards such as steepness and length of slope, presence of stones or rocks, risk of flooding, efficiency of the drainage, and vulnerability to wind erosion.

Recommendations as how the various parts of the land should be used will constitute the plan, which must recognize appropriate limits to rain-fed cropping as determined by rainfall, terrain, soils, and relationships with adjacent land uses such as forestry or grazing.

Studies of the relation between agriculture and climate, such as those carried out by WMO in Western Asia and Africa, have done much to determine the connections between climate and water needs of cereal crops, thus fixing the probability of the occurrence of effective seasons on the basis of climatic records.

Technical Measures (FAO[46]; SCS[45]; UNEP[17])

Contour plowing has been recommended for all lands with a slope of or exceeding 2%. It involves plowing on the contour, making a deep furrow every 30–40 m.

Crop rotation is a method to improve soil fertility and consequently increases soil resistance to erosion.

Mulching is practiced primarily to reduce sheet erosion. After the grain crop is harvested, the land is loosely plowed and the stubble left. The standing stubble serves to dissipate the strength of the wind and to provide a rugged surface to keep the soil in place (SCS[45]; FAO[46]).

Strip cropping is applied where contour plowing is not sufficient to control erosion. It comprises a strip of grain alternating with a strip left fallow or planted to summer crop. Usually strips are 20–30 m wide, with a deep furrow dug in the middle. The furrows store rain water.

Terracing is the most effective and widely practiced field measure for controlling or preventing erosion. It comprises series of mechanical barriers across the land slope to break the slope length and reduce the slope degree wherever necessary.

Checkdams are constructed in waterways originating in steep slopes. They reduce the velocity of water and arrest the silt which comes with runoff. Also, planting trees in the upper catchments and along gullies' margins and grassy areas that feed the gullies might be helpful.

Planting shrubs and trees as shelterbelts protect the land from wind erosion.

Efforts to rehabilitate degraded areas should form part of integrated action directed toward water management, improved land use, and the control of erosion.

The Extent of the Problem

According to the UNCOD[8] study, Dregne[40] estimated that 250 M ha is deteriorated rain-fed cropland. FAO,[46] using computer methods to define the magnitude of various variables including soil degradation, showed that about 25% (190 M ha) of total arable land in the developing countries need soil conservation measures.

FAO[38] estimated what may be a feasibility target to be attained by year 2000. The criteria of the feasibility are the availability of trained specialists, infrastructure, and the government priorities for soil conservation. This study considered a target of 25% as most feasible.

COMPLEMENTARY MEASURES FOR COMBATING DESERTIFICATION

Monitoring

There is a consequent need for regular monitoring of the status of the lands which have been immunized against desertification or improved to provide warning trends, identify areas in which change is taking place, and to provide a basis for investigation of causes and processes. It is in terms of such information that measures for prevention or reclamation will ultimately be designed. In the areas treated against salinization, alkalization, and waterlogging, samples of soils, irrigation water, and groundwater should be regularly analyzed to follow up changes in the salt concentration. The depth of the groundwater should be recorded. Indication plants should be observed and changes in the natural vegetation should be recorded.

A Global System of Monitoring Desertified Areas

The first recommendation of the Plan of Action to Combat Desertification[8] suggests that action programs should be based on assessment and evaluation of desertification and degradation processes leading to them. This means that "a system of survey and monitoring should be established (or strengthened) to assemble information on resources and populations and to carry out monitoring of the dynamics of desertification including the human conditions. The assembly and evaluation should be a continuous process, providing a feedback mechanism for national planning and action." The Plan of Action envisages national machineries for monitoring and assessment, regional or transnational cooperation in monitoring desertification processes of related natural resources (regional desertification monitoring centers), and a world-wide system for exchange of information gained from monitoring networks.

Monitoring methodology involves data acquisition at three levels:

- resource-sensing satellites,
- low-flying light aircrafts, and
- ground work.

Each method yields information of a particular type. The satellite method is initially the most important because it provides much information at a low cost over large areas for which descriptive and quantitative information may be entirely lacking.

The Global Environmental Monitoring System (GEMS) has initiated a transnational rangeland monitoring project in the Sahelian Region of West Africa (Price[47]).

LANDSAT imagery maps the static features, e.g., surface geology or inventory work (e.g., classification and identifying the patterns of rangeland vegetation resources). Images taken in different seasons or years are adequate provided there is at least one complete coverage (Price[47]).

When the purpose of monitoring is following changes in the ecosystem, it is required that imagery includes repetitive cover and must reach the user promptly.

Training

The personnel who are dealing with the irrigated lands should be well-trained to be able to carry out the agricultural processes correctly and efficiently, otherwise the soil will deteriorate. They should be capable of observing and interpreting changes in the land condition and the counteraction to be adopted.

It is always recommended that training courses in various disciplines involved in the project should be required before assuming the responsibility of any level in the project.

The farmers should be trained on practices of irrigated agriculture if they are not accustomed to these practices. The sensitivity of irrigated crops is such that unless the required process is carried out at the suitable time, the production decreases ending up with failure of the project.

In irrigated agriculture several types of machines and instruments are utilized. Efficient handling of these machines and instruments requires training of personnel. Workshops for maintaining and repairing the machines should be established.

Education

Successful attempts to control various features of desertification require a national consciousness of the soil conservation. A program of education should accompany or even precede the technical practices. The aim of this program is to supply the people with the necessary information and implication of the desertification processes. At the same time the program will enable the decision makers at various levels to recognize the dimensions of the problem.

National Institutions

Soil conservation on the national scale requires specialized institutions at national and local levels to plan and conduct the technical activities. The national council should be comprised of authoritative decision-makers and top technical consultants so that the decisions made are carried through to action.

Legislation

It is necessary to legislate implementation of programs of soil and water conservation. Several countries have such legislation. However, they need to be received within the framework of overall scientific land-use planning and development. Legislation should state clearly the powers and responsibilities of different authorities charged with implementing the laws, the rights and responsibilities of the land owners, the provisions of financial contributions, subsidies, credit facilities, and financial resources.

Incentives

In order to raise people's interest in the control program, incentives of various kinds should be provided and publicized. It is of importance that the people cooperate with the implementation of the soil and water conservation program.

SOIL EROSION

Soil erosion refers to the removal, transportation, and net loss of soil. It involves the loss of soil fertility as well.

Normal and Accelerated Erosion

Erosion which takes place without the influence of man by the action of wind, water, temperature changes, and biological activity is considered as natural or geological erosion. The loss of soil by this natural phenomenon is regenerated when the soil has a dense vegetation cover, by natural means at the same rate as it is removed. Thus the soil and its vegetative cover are in a state of balance. This erosion is termed "normal erosion" (FAO[46]).

When the state of balance between the soil and its vegetation cover is disturbed by cultivation, grazing, or burning, the soil becomes exposed to direct wind and/or water action. Thus the soil can be washed by water or blown by wind at a faster rate than it can regenerate, resulting in a net loss of soil. This erosion is termed "accelerated" (FAO[46]).

Deforestation, plowing the pasture land and eliminating the natural vegetative cover and leaving the soil surface bare, leads to soil erosion with wind or water and the transformation of the productive soil to unproductive or less productive which decreases the national income. The problem of soil erosion is one of the most serious problems from which farmers of various regions suffer.

The mechanism of desertification in the case of dust storms, forageland, and rain-fed cropland is the soil erosion by wind or water.

Soil erosion with wind or water affects the soil in various ways.

Loss of Soil Fertility

At the beginning, the surface fine soil particles are detached and moved under the water or wind action. Because this soil layer is rich in nutrient elements, the eroded soil thus loses its fertility. The average yearly loss of the cultivated soil may be as much as 300 tons per hectare per year. This amount equals the weight of a soil layer 2.5 cm deep for an area of one hectare. Accordingly, after 6–7 years the plowed layer may be lost by erosion. Generally yearly loss of 10 tons/ha is considered serious. Plowing the eroded land does not solve the problem because the soil plow-layer in this case is the subsoil layer which is less fertile. Nutrient elements in this layer are not in forms available to plants. Also, the biological activity in this layer is weak (SCS[45]; FAO[48]; Lal[49]).

Decrease of Cultivated Land

The eroding processes do not affect the surface layer only but also affect deeper soil layers, until the parent material is reached. This has been clearly noticeable in coastal slopes which results in uncultivable soil. Trees may be easily pulled out from the eroded soil due to the loss of the soil of the root zone. The cultivated soils are subject to the flow of the eroded materials moved by wind or water which destroys the growing plants. Blowing sand may cover the cultivated areas adjacent to the desert and damage the plants. The net result is decrease of the cultivable area.

Damaging the Constructions

The eroded materials moved with the running water or the blowing wind may bury water canals and drains due to the deposition of these materials. Also, reservoirs may be damaged by silting, thus storing capacity decreases.

The Basic Steps of Soil Erosion

Soil erosion by water or wind takes place in two steps. The first is breakdown soil aggregates and detaching soil particles (Lal[49]). The disintegration of soil clods or the dispersion of coagulated particles in the soil surface layer affects their mechanical stability (of the soil particles). This disintegration process depends on the ability of the eroding agents to destroy the materials responsible of cementing the soil particles and binding them into soil clods. Accordingly, soils having fine texture well-coagulated particles are less subject to erosion and transport than the sandy soils made of single particles which require low energy to detach them.

The second step is soil transport (Lal[49]). The transport of the soil material is a real loss. The detached soil particles have small size and weight and thus are easier to transport than the large particles which have larger size and weight.

The wind and running water movement generates the energy required to disintegrate and transport the soil particles. Rainfall does not cause soil particles entrainment unless there is a flow of water on the soil surface. This takes place when the rainfall exceeds the infiltrating water through the soil or when the soil is sloping. The ability of wind to transport the soil particles depends on the soil slope, soil characteristics, and soil utilization.

WATER EROSION

Water erosion is caused by various sources of water: rainfall, melted ice, irrigation water, rivers, and other water courses. Rainfall erosion is more spread out than other causes.

The rainfall characteristics of importance in soil detachment are drop size distribution and the angle and velocity of its impact. The latter depends on whether the rain is wind-driven or not. Actually, kinetic energy of the rain is the principal factor responsible for soil detachment, although momentum is also considered important (Lal[49]).

Rainfall causes soil erosion when rain water flows on the soil surface. In the case of arable lands, erosion takes place when the soil surface is barren without protection by vegetation cover. Erosion also takes place when water infiltration through the soil decreases. Most of the programs to combat soil erosion by water are based on improving water infiltration through the soil, thus water flow on the soil surface decreases, and consequently soil erosion decreases.

When decreasing the water flow on the soil surface is not feasible, other means to impede its movement are applied to control water erosion. Generally, dense soil vegetative cover is considered most effective in protecting the soil from water erosion. It does not protect the soil from the collision of the water drops with the soil surface only, but also impedes the water flow on the sloping soil.

Forms of Water Erosion (Lal[49]; Soil Survey Staff[50])

Splash Erosion

When the drops of rain water fall on the soil surface, each drop collides with the soil and bursts as a bomb shell. Thus,

- The soil clods are detached into small single particles.
- Upon the rain drop collision with the soil surface, it is fragmented and splashed, carrying the single soil particles far from the spot of collision downward the slope more than upward. The net movement of water is usually downward the slope, Soil Survey Staff.[50] This form of water erosion is most important and serious.

Sheet Erosion

This form of water erosion takes place when:

- the disintegration of the soil clods and their movement are regular,
- the soil is directly affected by heavy rain,
- the soil surface is smooth and regularly sloping, and

- the rate of rainfall is more than the rate of water infiltration through the soil, thus water accumulates on the soil surface and starts to flow to the lower areas.

The water flows above the soil surface as a cover having a regular thickness. This form is usually slow and is not noticeable and requires measurement.

Rill Erosion

When rain falls on soils having regular slope, the soil holes and cracks are filled with water. The water starts to flow upon the increase of its amount until it spills over the sides of holes and cracks causing erosion of the sides and bottoms of these channels. These small channels are combined, forming larger but flat channels usually termed rills (SCS[45]; FAO[48]).

Gully Erosion

The water runs on the soil surface in small channels toward the major slopes. Its movement and ability to erode the soil is increased, forming gullies (SCS[45]).

The gullies magnify the destructive power of the water because it is concentrated in deep channels and moves faster and more forcefully than when it moves in thin layers along the slope. Because the water in the gullies contacts the least area of land for a short period of time, the infiltration rate of water in the gullies is low which increases the water flow and its ability to erode the soil. Although gully erosion is important, the sheet and rill forms of erosion are more frequent and more serious.

Factors Affecting Water Erosion (Lal[49])

Water erosion is affected by several factors:

- Rainfall quantity, frequency, and distribution among the year seasons.
- Gradient and length of soil slope.
- Kind of vegetative cover, its residues, and proportion of the area covered by plants.
- The culture practices which decrease the surface flow.
- Water runoff in arid regions is characterized by flash floods having a high velocity but short duration. This is due to the nature of rain which falls in short torrents and to the low water infiltration through the soil.
- The soil properties which affect the water infiltration such as particle size distribution, structural stability, and crusting.

Thus under arid ecosystems (Lal[49]):

- Entisols that are derived from recent deposits and have sand texture are easily eroded.
- Aridisols contain low organic matter. Desert pavement on its surface indicates past erosion of fine soil particles.
- Mollisols contain relatively high organic matter, thus they are less eroded.
- Alfisols contain well-defined argillic horizon. They are highly erodible by water.

- Vertisols contain considerable clay proportion, thus they have low infiltrability and high runoff. They form larger and deep cracks and are highly susceptible to erosion.

Evaluation of Soil Erosion by Water

Visual Evaluation

This method is qualitative and based on reconnaissance surveys. Soil Survey Staff[50] has developed a rating system to assess the degree of soil erosion as follows:

Classes of Erosion by Water (Soil Survey Staff[50])

Class 1: The soil has a few rills or places with thin A horizons that give evidence of accelerated erosion, but not to an extent to alter greatly the thickness and character of the A horizon. Up to about 25% of the original A horizon or original plowed layer in soils with thin A horizons may have been removed from most of the area.

Class 2: The soil has been eroded to the extent that ordinary tillage implements reach through the remaining A horizon or well below the depth of the original plowed layer in soils with thin A horizons. Generally, the plow layer consists of a mixture of the original A horizons and underlying horizons.

Class 3: The soil has been eroded to the extent that all or practically all of the original surface soil, or A horizon, has been removed. The plow layer consists of materials from the B or other underlying horizons. More than about 75% of the original surface soil, or A horizon and commonly part or all B horizon or other underlying horizons, have been lost from most of the area. Shallow gullies, or a few deep ones, are common on some soil types.

Class 4: The land has eroded until it has an intricate pattern of moderately deep or deep gullies. Soil profiles have been destroyed except in small areas between gullies. Such land is not useful for crops in its present condition.

Table 41 presents an evaluation of the soil and climatic factors that affect erosion by water and a classification of erosion categories (Lal[51]).

Method of Direct Measurement: Experimental Field Plots

The information obtained from this method was the base upon which it was possible to express erosion by the Universal Soil Loss Equation (USLE).

The plots are rectangular areas perpendicular to the contour lines. Data about soil erosion and the factors which control it are collected in the field, though simulated plots and conditions may be established in the laboratory. Field data are usually more factual. However, the changing of conditions in the field makes the assessment of the main factors of soil erosion difficult.

Experiments are usually conducted to assess the effect of one factor with the other factors being as constant as possible. Experiments to study the effect of slope should be carried out on soils having variable slopes. Also, the length of the slopes has to be 5–10 m. Replicates are necessary to obtain satisfactory results.

Table 41.   Soil and climatic parameters that affect erosion by water.

Erosion Category	Depth (cm)	Org. C. (%)	Erodibility (K)	Slope (%)	Erosivity (R) (foot-ton/acre per ammum)
Very serious	<10	0.5–1.0	0.6	>10	>1000
Serious	10–20	1.0–1.25	0.2–0.6	5–10	600–1000
Moderate	20–50	1.25–2.0	0.05–0.2	2–5	400–600
None to slight	>50	>2.0	<0.05	<1	<400

Source: Lal.[49]

The standard experimental plots are 22 × 1.8 m though other units may be 5 × 20 m or 2.5 × 50 m. The edges of the plots are made of wooden or metal strips or any other impermeable noncorrosive material. The edges are about 15–20 cm above ground level. At the end of the slope the runoff water is collected in a covered basin. The collected water is directed through a canal to collecting stores. A flocculating agent is added to the mixture of water and sediments. Thus, the soil material is deposited in the bottom. The clear water is drawn and measured. Also, the volume of the soil material is determined and sampled to determine its volume and weight. The total weight of the deposited soil material can thus be calculated. This weight is the amount of eroded soil from the plot if only one collector is used to receive the runoff water from the plot (Nahal and Darmash[52]).

The rainfall is measured using a standard automatic recorder close to the experimental units.

Comparison of Eroded Soil with Uneroded Adjacent Soil

The comparison may be carried out to evaluate the erosion which had taken place in the surface layer of the eroded soil as compared with the uneroded soil profile due to the presence of vegetation cover.

Another method for comparison was suggested by Ritchie et al.[53] using caesium (^{137}Cs).

Because ^{137}Cs fall-out takes place after each nuclear experiment since the 1950's and the sixties, it is adsorbed on the soil colloidal surfaces in the whole world Accordingly, comparison of the content of ^{137}Cs in a site subjected to erosion with another adjacent site which was not eroded gives an estimation of the erosion which had taken place during the last thirty years, as erosion results in decrease of ^{137}Cs in the eroded area.

Evaluating Sediments in Front of Dams

Dams are constructed in watersheds. The deposits brought by water accumulate in front of the dam and cause silting of the reservoir. Evaluation of these deposits gives an estimate of the amount of the eroded soil (Le Houerou[11]).

Evaluating the Factors Controlling Water Erosion

Researchers were not able to quantitatively express the soil loss by water erosion until the agricultural experimental stations had conducted numerous experimental plots described above and unlimited results describing the contribution of each factor involved in water erosion phenomenon. Thus researchers ended up with the "Universal Soil Loss Equation" (USLE). The same equation is also used to assess the needs for soil conservation to combat desertification (SCS[45]). The equation appears as follows (Wischmeier et al.[54]):

$$A = R.K.L.S.C.P. \tag{40}$$

where A = the average calculated amount of lost soil for each area of land per year (t/acre).

R = rainfall erosive index or erosivity of rainfall. It is an evaluation of rainfall ability to erode soil.

K = erodibility factor of the soil. It is the erosion average for the unit of rainfall erosion index for a specified bare soil without any conservation treatment. Its slope is 98 and length 22 m.

L = the land slope length factor. It is the ratio between the amount of lost soil from a field having a specified slope to the lost soil amount from a field plot having a length of 22 m and have the same slope.

S = the slope factor. It is the ratio between the loss from a specified soil and the loss from a soil of the same type having a slope of 9% and the same slope length.

C = is the cropping management factor. It is the ratio between the amount of loss of a specified soil having a management system planted to specified crops, and the soil of the same type having the same "K" value and without crops.

P = is the soil conservation practices factor. It is the ratio between the loss from a soil with practices of conservation to the loss of soil without these practices.

The "USLE" estimates the average annual soil loss. It does not predict storm-by-storm soil loss. According to Goudie[2] it does not apply under the following conditions:

- Regions where the equation terms are not available.
- Complex watersheds.
- Slopes steeper than 20%.

Also, rainy storms of less than 1.25 cm of rain cannot be taken into consideration. The yearly rainfall erosive index is calculated as the summation of the indices of all raining storms during the year. The index used in "USLE" is the average of indices for several years.

The rainfall erosive index "R" can be calculated from the following (Wischmeier et al.[54]):

$$R = E\,I_{30}\,/\,100 \tag{41}$$

where R   is the rainfall erosive index.

E   is the total kinetic energy of rain.

I_{30} is the highest rain intensity in 30 min. Rain intensity is related to the duration and recurrence interval or return period of the storm.

Other investigators (Hudson[55]; Fournier[56]; FAO[57]; Foster and Lane[58]) had suggested other empirical formulas for estimating erosivity by rain (Table 42).

Because USLE factors are developed from experiments performed on cultivated areas, the value of C, the crop management factor, needs to be determined. A subfactor approach was proposed by Wischmeier,[59] Mutchler et al.,[60] and Laflen et al.,[61] as stated by Renard.[62] Thus, the factor C is expressed as:

$$C = LU \cdot CC \cdot SC \cdot SR \tag{42}$$

where   LU is a land surface subfactor.
   CC is a canopy subfactor.
   SC is a surface cover subfactor.
   SR is a surface roughness subfactor.

Renard[62] proposed subfactors for rangeland as follows:

$$LU = 0.40 \cdot \exp(-0.12 \cdot RS) \tag{43}$$

where RS is the live roots and buried residue in the upper 100 mm of soil (kg per ha per min of depth).

Because this number is not exactly easily obtained, a scheme has been developed to estimate the value from annual above ground biomass estimates:

$$RS = B10 \cdot n \cdot 2/100 \tag{44}$$

where   B10   is the annual above ground biomass estimate (kg Der ha).
   e   is the ratio of below ground biomass to above ground biomass.
   n   is the ratio of biomass in the upper 100 mm of soil to the total below ground biomass.

Tables for grass brush, trees, etc., are being prepared as stated by Renard[62] in February 1987.

$$cc = 1 - fc \cdot \exp(-0.34 \cdot H) \tag{45}$$

where   FC is the fraction of the land surface beneath canopy.
   H   is the height (m) that rain drops fall after impacting the canopy.

$$SC = \exp(-4.0 \cdot M) \tag{46}$$

where M is the fraction of the land surface covered by nonerodible material (e.g., living and dead plant material, rock and large gravel). This factor is important especially where erosion pavement, cryptograms, or other nonerodible items can be expected to protect bare soil from the erosive forces of raindrops or flowing water or both.

$$SR = \exp(-0.026 \, CRB - 6)(1 - \exp[-0.035 \cdot RS]) \tag{47}$$

where RB is a random roughness (mm) expressed as standard deviation of surface elevation from a plane and is intended to reflect any tillage consequence or other roughness forms.

Table 42.   Empirical formulas for estimating erosivity by rainfall.

Formula	Description	Reference
EI_{30}	E = Kinetic energy of rain, I_{30} = maximum 30 minimum intensity	Wischmeier et al.[54]
KE > 1	Kinetic energy of rains with intensity exceeding 25 mm/h	Hudson[55]
$C = p^2/P$	C = climatic coefficient p = mean monthly rainfall during wettest month of the year P = annual rainfall	
$^{12},p^2/p$	modified fourniers index	FAO[48]
AI_m	A = amount of rainfall per storm (cm) 1_m = max. intensity over 7.5 min. time interval	Lal[49]

Source: Lal.[51]

At the end of an article entitled "Beyond the USLE," Foster and Lane[58] concluded that with the development of powerful and readily available personal computers, the USLE will be replaced with erosion prediction technology based on fundamental hydrologic and erosion processes. Realizing this goal requires development in the computer technology as well as determining various parameters of erosion processes, they stated that the USDA-Water Erosion Prediction Project (WEPP) is expected to represent a major turning point in erosion research.

Combating Water Erosion (Lal[49,51])

Erosion of soil by water causes great losses. The soil should be conserved so that it can resist water erosion and subsequent desertification.

Soil conservation means that the soil content of organic matter, nutrients, and applied fertilizers are not lost. It also means that low-lying soil areas are protected from being covered with the eroded sediments from the higher level soils and protecting water courses from being silted by these deposits.

As stated above, the rainfall erosivity as well as the soil characteristics and its vulnerability to erosion cannot be changed. However, the coarse-textured soils are more easily eroded by water than the fine-textured soils.

The remaining factor that affects water erosion is management. Through this factor it is possible to control soil erosion with water. This can be achieved through two main approaches (FAO[46]; Nahal and Darmash[52]):

- Kind of crops and their rotation which is termed "Biological erosion control," and
- The control of water movement termed "Mechanical erosion control."

Biological Control of Water Erosion

Land Use

The base of soil conservation is identifying the problem. Growing grains on the steep slopes of mountains and shallow rooting depths requires terracing, but constructing permanent effective terraces is so expensive that growing cereals would not be economical. Accordingly, the cereals do not suit such sloping areas. Rather, pasture is more suitable and more profitable.

Crop Management

Crop management is the most effective approach to control soil erosion by water. Terracing is known to decrease soil erosion to 50%, but crop management may decrease it to 10%. The aim of crop management is to prevent direct contact of water drops with the soil surface to decrease the mechanical action of rainfall drops, thus erosion may be decreased. This can be achieved by keeping the soil surface covered with the natural vegetation or by crops. Also, growing the crops in rows has a significant effect because the plant population is increased in the field.

Diverse farming systems—a mix of crops, trees, and livestock—are more effective in minimizing soil erosion risks than intensive cropping systems based on monoculture and overstocking the area with animals (Lal[51]).

Also, selection of crops is important. Grain crops should be grown in rotation with fallow planted by leguminous cover or lightly grazed pasture. Mixed cropping of cereals and legumes is more effective as stated above (Lal[51]).

An objective of controlling water erosion by crop management is to decrease splash erosion through providing effective protection for the vegetative cover. Thus any technique to provide this cover and improve its growth by fertilization and other means is recommended.

However, in the periods which precede the crop germination and/or follows harvesting the crop, the soil is bare and subject to water erosion. To protect the soil in these periods the soil surface should be mulched with crop residues.

Mechanical Control of Water Erosion

The mechanical practices to control the soil erosion with water are expensive. Thus it is recommended to apply them in productive lands because these lands are more valuable and are subject to water erosion.

The aims of the mechanical practices may be one or more of the following:

a. Impeding the water movement.
b. Increasing the soil ability to retain and store water.
c. To safely get rid of excess water.

The mechanical means to control soil erosion with water depend on the cropping pattern. Among these measures are the following.

Contour Farming

Contour tillage, plowing parallel to the contour lines, is practiced on lands having slight slope (Lal[51]; FAO[46]). Thus, water runoff slows down and water is directed to the furrows before it runs downward. Thus more water penetrates through the soil resulting in decreasing the soil erosion. Complementary to the biological control of water erosion are practices that improve seed germination. Among these practices are:

- Tillage improves water infiltration through the soil as it breaks the surface crust and the impermeable hardpans which may be encountered in the soil profile. Accordingly, the amount of water runoff is decreased as well as germination of seeds is improved (Lal[51]).
- It has been stated above that mulching protects the soil surface.
- Fertilization improves the plant growth, thus a better vegetative cover is attained.
- Soil practices that improve its physical properties will be reflected on seed germination and plant growth and its efficiency in protecting the soil surface.

Grass Strips (FAO[46])

This practice is usually applied in rangeland. The sloping area is divided into horizontal strips parallel to the contour lines. Crops are planted in the strips in exchange with the natural vegetation cover. When rain water flows on the sloping land, it slows down. The width of the grass strip depends on the slope of the land. It increases with the increase of the slope, while the width of the crop decreases with the increase in the slope.

Contour Bunds (FAO[46])

Soil bunds 1.5–2 m wide, perpendicular to slope of the land to impede the water flow, increase water infiltration through the soil. The bunds are usually 10–20 cm apart. They are effective in lands having slight slopes only.

Ridging and Tied-Ridging

Form small channels parallel to contour lines as a part of the farrowing process. Each channel has a slope so that the water it receives can be collected far from the protected land.

The tied-ridging is the formation of a network of small channels, thus the surface of the land is divided into small basins storing the water and allowing it to penetrate through the soil. This system is effective in the case of deep permeable soils.

Ridges and Furrows

The land is plowed to form furrows 10–15 m apart with channels 1/2 m deep. The surface water flows above the furrow and is received in the slope of the channel. This system is more suitable for the large areas having slight slopes but needs surface water control.

Terraces

The terrace is a flat wide furrow of soil, perpendicular to the slope of the land to cut the surface runoff and leads it to an outlet at suitable speed. Also, terraces are used to shorten the gradient. They differ in length and width and arrangement. Lal[49] reviewed water erosion in Goudie.[2] He stated that according to Nicou,[63] tillage that provides a rough cloddy soil surface is more effective in controlling runoff than tillage which makes the soil surface smooth and devoid of vegetation cover. In this regard discing and harrowing should be avoided as shown by experiments in Tunisia (Floret and Floch[64]). The same conclusion applies to excessive tillage. Arab Center for Studies of Arid Zones and Dry Lands[65] (ACSAD) recommends the following to control erosion in arid regions by tillage:

- Contour plowing with deep, steep-walled furrows to retain heavy runoff.
- Economical crops having extensive root system are planted on the furrows.
- Rill formation should be controlled.

Diversion Terraces

These terraces are made to impede runoff water on the soil surface and to divert it across the slope to a suitable outlet. The diversion terraces are not suitable for lands with slopes more than 7° as their construction will be expensive.

Retention Terraces

These terraces are utilized where storing water at the sides of the hill is necessary. They are applicable only in permeable soils with slopes less than 4.5°.

Bench Terraces

These terraces are made up of a series of benches and soil ridges. The ridges are strengthened against the water flow by growing dense plants or by stones and bricks. In these terraces no channels are made but an area is used to store the water. They are usually more suitable under the following conditions:

- Available inexpensive labor.
- Where high priced crops are grown.
- Heavily populated areas where it is necessary to utilize every hectare to produce food.

Waterways

The function of the waterways in the conservation system is to control the surface water and carry it before it has the time to erode the soil. There are three kinds of waterways:

a. Diversion channels constructed in the elevated part of the sloping land to receive the water flowing downward and divert it across the slope to the grass waterways.
b. Terrace channels are also constructed to collect the surface water. They differ from the diversion channels as they collect the water flowing from the upper area.
c. The grass waterways function is to receive the water collected from the other channels and discharge it into a natural river. These channels suit slopes up to 11° and should be fortified by stones if the stone reaches 15°. Plants are grown in these channels to establish a dense cover.

Gully Erosion Control (SCS[45])

Gully erosion takes place usually in uncultivated soils. The control of this type of erosion is costly and difficult. There are several factors that should be taken into account before a decision to carry out this task is taken. The reasons which cause the formation of gullies should be studied so that counteractions can be taken.

There are several ways to control gully erosion.

Check dams of various types and diversion channels are prerequisites for controlling gully erosion.

Check dams are successions of small dams along the slope of the area to impede water movement, thus deposition of solids and infiltration of water through the soil increase.

Rapid gullies may require other types of structures such as "drop structures."

According to Lal[49] there are two kinds of drop structures. The low drop structure is a drop in which the bed drop height is equal to or less than the upstream specific head, that is less than the upstream flow depth plus the velocity head. High drop structures are those in which the upstream water levels are normally unaffected by downstream conditions, such as high tailwater levels that submerge the spillway crest.

Gully Erosion Control by Vegetation (SCS[45])

Gully erosion with vegetation is preferred because wooden or cement structures are subject to erosion and disintegration while plants grow and flourish over time. The plants slow the water movement and increase silt deposition. Silt sediments improve the plant growth. The difficulty of the vegetative control is that the gully soils are usually poor in plant nutrients, structureless, do not contain organic matter, and their water-holding capacity is low. These difficulties are confronted by selecting the suitable plants, utilizing special agricultural practices.

To help the vegetation grow, the following techniques may be followed:

* Healthy seedlings in polyethylene bags without a bottom filled with fertile soil and laid into holes made with an auger. Over time the plant will grow out of the bag.
* Colonies of plants in bags filled with fertile soil laid in channels at the same level of the gully bottom. The bags protect the soil removed by water current at the first flood.
* The sides of the gully may be planted after leveling. The soil of the gully sides needs to be fertilized or mixed with fertile soils. The planted sides can also be mulched to improve germination and protect the seedlings.

Temporary Structures

When the erosion is so strong that it is difficult for plants to grow, or when the soil is so infertile, temporary structures to protect the plants of the following kinds are made.

Wire Blosters

A galvanized wire net 2 m wide is laid on the gully bottom. Fragments of stones are piled on half of the wire net width. The other is wrapped over the stones and tied.

Netting Dams

The wire net can be used to form a small dam. Plant residues and straw are laid at the net side facing the water flow, thus making a porous barrier, which slows the water current.

Log Dams

Two rows of wood logs are buried vertically in the gully bottom. Their height should exceed the height of water. Other logs are arranged in between.

Brick Weirs

Bricks may be used to build a dam until the plants are established.

Permanent Structures

Several types of structures are used to control gully erosion such as silt trap dams, regulating dams, gully head dams, and drop structures.

Symptoms of Water Erosion

The symptoms which indicate that soil was eroded by water may be noticed especially a short period after the storm. Among these symptoms are the following (Soil Survey Staff[50]):

- Presence of rills or small gullies, especially in the upper slopes or the sides of roads.
- The flowing water has a clayey appearance. This is especially clear during the storm or shortly after.
- The formation of various degrees of gullies is a sign of the erosion problem.
- In the cultivated fields, the presence of uneroded soil pillars is a sign that erosion has taken place around these pillars.
- Presence of pebbles and stones on the soil surface after eroding the fine particles.
- Accumulation of heaps of soils and debris.
- Accumulations of soil material surrounding the trees in the slopes.
- Trees' roots are nuded.
- Light-colored circles around the rocks and lichens lines.
- Soil materials sedimentation on the medium slopes.
- The soil parent rock is eroded and nuded.
- Slides of soil and rocks at the end of the slope.
- The barren spots in the rangeland regions result from overgrazing. They indicate that the soil was eroded or in the process of erosion.
- Unlevel soil surface and light-colored and deep-colored soil surface spots.
- Sedimenting of gravel, sand, and silt in the gully beds.
- Plant roots in the streamlets are denuded.
- Change of the natural plants.
- Silting of water reservoirs.

Classification of water erosion intensity introduced by FAO[7] was presented above. Other systems of water- and wind-erosion classification to "slight," "moderate," "severe" and "excessive" grades were suggested (UNCOD Secretariat[14]; Soil Survey Staff[50]).

WIND EROSION

Erosion of soil by wind is an important problem in arid ecosystems where the wind has a high velocity, the soil clods are easily broken down and entrained, and the vegetation is meager and provides only a cover for a slight proportion of the soil surface.

According to Armbrust,[66] wind erosion can occur whenever (1) the soil is loose, dry, and finely divided, (2) the soil surface is smooth and vegetative cover is sparse or nonexistent, (3) the field is sufficiently large, and (4) the wind velocity is high enough to move soil.

Mechanisms of Wind Erosion

Wind erosion occurs due to two mechanisms:

- Breakdown of soil clods and aggregates into single particles.
- Transportation of these particles by wind energy.

The more wind load of soil particles, the greater is its ability to breakdown and move more soil particles. The movement of soil particles occurs according to one or more of the following processes (Constantinesco[46]; Soil Survey Staff[50]).

Saltation

The forward movement of wind occurs in a spin-like form and usually in strong successive waves. The soil particles jump during their movement as they go up and drop several times. The collision of the dropping particles with the soil surface increases their effect on soil aggregates. The circulating movement of wind also increases its ability to detach the soil particles and entrain them. The soil particles' saltation is considered one of the most effective mechanisms of wind erosion. Soil particles having diameters within the range of 0.05 to 0.5 mm move in saltation.

Creeping

Winds of slow speed may be unable to uplift soil particles of large sizes, thus these particles may creep on the soil surface. They also may collide with other particles, thus moving them. Winds are not able to carry soil particles if the velocity of their layer contacting the soil surface is less than 12 knot per hour (Fryberger[68]).

Transportation in Suspension

Small particles of fine sands or smaller move by uplifting in the air in the form of an air suspension. They remain suspended in the air and are not deposited on the land before the wind settles or rain falls.

Factors Affecting Wind Erosion

Climatic Factor

The relationship between the climatic conditions and wind erosion depends on the intensity and recurrence of wind, rainfall, evaporation, and the soil surface moisture during wind blowings. The moist soil is usually less movable than the dry soil. When the soil moisture reaches a lower percentage than the wilting point, wind starts to erode the soil and its ability increases with the increase in its velocity (Figure 22).

Wind is a main factor in moving the soil particles by virtue of its energy. Several factors are involved in the relation between the ability to transport soil particles and various atmospheric variables, among which are wind velocity, turbulence, gustiness, shear forces, humidity, and temperature (Middleton[67]).

The Soil Factor

The erodibility of soil by wind depends on the detachability of single soil particles or aggregates from the soil and on the size of these particles and aggregates. The smaller the soil particles the easier to move with wind.

According to Chepil and Woodruff,[28] soil erodibility depends largely on its mechanical stability which they defined as "the resistance of dry soil to breakdown by a mechanical agent such as tillage, force of wind or abrasion from wind blown material." This mechanical stability depends on size, density, and shape of the soil individual particles.

The sandy soils are made up of single particles, thus the size of the particles is the factor that affects the ease of entrainment. The large sand particles do not move with wind because of their weight. When the particle diameter decreases toward 1 mm, its mobility increases. With further decreases to less than 0.1 mm, attraction between particles increases until the particles become resistant to movement and the breakdown becomes more difficult, thus the erosion decreases.

The Roughness Factor

Soil roughness is the result of plowing the fine-textured soils or farrowing the coarse-textured soils using agricultural machines. It depends on the tillage implement.

Soil roughness increases soil resistance to wind erosion by catching the soil particles suspended in the air behind the furrows' ridges.

The waves of the soil surface catch the soil particles regardless of the wind direction while the furrows have the same effect if their angle suits the wind direction.

According to Fryreer,[27]

$$\text{Ridge roughness} = \frac{4 \times \text{field ridge height in inches}}{\text{field ridge spacing/field ridge height}}$$

He compared several tillage implements according to estimated soil roughness produced (Table 43, Figure 23).

Wind erosion occurs mainly right after seeding because the culture practices to prepare the seed bed leave the soil surface smooth and powdered, thus easily eroded. Seeding in the bottom of furrows protect the soil from wind erosion.

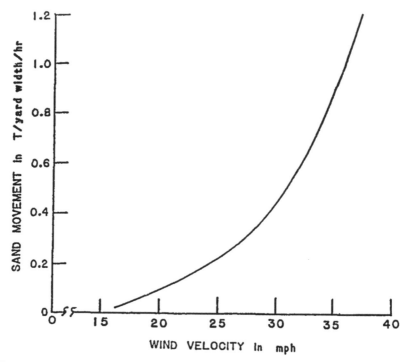

Figure 22. Relation between wind velocity and rate of sand movement (Fryreer[27]).

Vegetation Cover

The growing plants as well as plant residues in the field help to protect the soil from wind erosion. They decrease the wind velocity at the soil surface and protect areas with loose particles (Figure 24, Table 44).

Growing plants or mulching should cover the soil surface as completely as possible in areas subject to strong winds. Growing plants are 2–2.5 times as effective as mulching with plant residues in protecting the soil from wind erosion.

Field Width

Increasing the number of entrainment and collision of the soil particles with the soil surface, their ability to detach more soil particles increases, thus wind erosion increases. Accordingly, the total amount of soil entrained by wind depends on its movement distance across the field or on the width of the field. For each region according to its climate and soil, there is a maximum field width beyond which wind erosion does not significantly increase. In regions having high erodibility with wind, the amount of soil transported by wind reaches its maximum after a relatively short distance. Thus unprotected regions, easily erodible, should be as narrow as possible (Armbrust[66]).

Strip agriculture is considered among the most suitable practices for determining the field width which faces the wind. The strips are arranged to face the wind direction. Strip agriculture is considered a common practice in arid ecosystems. Strips of all cereals are planted in exchange with strips left bare. The bare strips usually store moisture.

Armbrust[66] summarized the primary variables that control wind erodibility as follows:

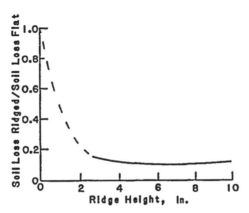

Figure 23. Influence of soil ridge height on wind-eroded soil losses (Fryreer[27]).

- Soil Erodibility Index, I, the potential soil loss in tons per acre per year from a wide, unsheltered, isolated field with a bare, smooth, noncrusted surface under the climatic conditions in the vicinity of Garden City, Kansas.
- Knoll Erodibility Stability, I_s, accounts for the effect of windward slopes of less than 500 feet if a knoll exists in the field.
- Surface Crust Stability, F_s, is the mechanical stability of the crust.
- Soil Ridge Roughness, K_r, is the roughness of the soil surface caused by ridges and furrows.
- Velocity of Erosive Wind, V, is the mean average wind speed corrected to 30 feet.
- Soil Surface Moisture, M, accounts for the resistance of soil moisture on soil particle movement.
- Distance Across Field, D_f, is the distance across the field measured along the prevailing wind erosion direction.
- Sheltered Distance, D_b, is the distance along the prevailing wind erosion direction that is sheltered by a barrier.
- Quantity of Vegetative Cover, R'_s, is the amount of surface crop residue in pounds per acre.
- Kind of Vegetative Cover, K_s, is the factor describing the total cross-sectional area of the vegetative material.
- Orientation of Vegetative Cover, K_o, is the vegetative roughness factor.

The wind erosion equation (Armbrust[66]; Lal[49]) combines I and I_s into I; V and M into C; D_t and D_b into L; and R'_s, S, and K_o into V. The factor F_s is ignored and K_r becomes K. Thus, potential average annual soil loss by wind erosion "E" is a function of

$$E = f(I, K, C, L, V) \qquad (48)$$

where I = soil erodibility
K = ridge roughness factor
C = local climatic factor
L = field length along the prevailing wind erosion direction
V = equivalent quantity of vegetative cover.

The information needed to solve Equation 48 for a particular field are:

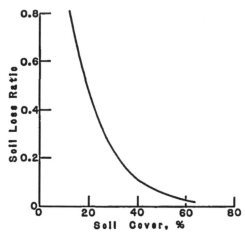

Figure 24. Influence of soil cover on soil loss ratio. Soil loss ratio is soil loss with various percentages of cover divided by soil loss from bare soil for same wind velocity (Fryreer[27]).

Table 43.   Tillage implements ranked according to estimated soil roughness produced.

Implement	Ridge Roughness[1] (inches)
Lister	6–10
Chisel	3–6
Planters or drills	2.5–6
Small sweeps	2–6
Large sweeps	1.5–4.0
Moldboard plow	0–1.5
Disk plows	0–1.5
Subsurface rod weeder	0

[1]

$$\text{Ridge roughness} = \frac{4\times \text{ field ridge height in inches}}{\text{field ridge spacing/field ridge height}}$$

Source: Fryreer.[27]

- percent of aggregate larger than 0.84 mm in diameter,
- height and spacing of ridges,
- width of the field,
- height of barrier if present, and
- kind, orientation and amount of residue.

Combating Wind Erosion

Practices applied to combat sand movement presented above are the same measures to control wind erosion. These principles may be summarized into the following (Fryreer[27]):

Table 44.   Crop survival as influenced by duration of exposure to a 30 mph wind with sand flux of 1.0 ton/rod width/m on plants 9 or 10 days old.

Crop	Survival Rates at Three Exposure Times		
	5	10	20
		%	
Pepper	75	8	0
Onion	100	100	100
Cabbage	100	87	56
Southern pea	100	94	72
Carrot	91	10	4
Cucumber	100	100	46
Cotton	100	85	15
Sunflower	91	88	72
Avg.	95	72	46

Source: Fryreer.[27]

- The soil surface should be covered.
- Increasing the soil surface roughness.
- Reducing the wind velocity.

He presented Tables 43 and 44 and Figures 22, 23, and 24 to show the efficiency of applying these principles in controlling wind erosion. The aggresivity of wind "C" is estimated by an empirical relation that involves wind velocity "V" and precipitation effectiveness of Thornthwaite as follows (Lal[51]):

$$C = V^2 / 2.9 \ (PE)^2 \tag{49}$$

Classes of Erosion by Wind

The Soil Survey Staff[50] had classified the soil erosion by wind in the following classes.

Class 1: Wind has removed from the soil a sufficient amount of the A horizon that ordinary tillage will bring up and mix the B horizon or other lower lying horizons with surface soil in the plow layer. Generally, about 25 to 75% of the original A horizon (or surface soil in soils with thin A horizon) may have been removed.

Class 2: Wind has removed all of the A horizon and a part of the B or other lower lying horizon. The plow layer consists mainly of the original horizons below the A horizon.

Class 3 (Blown-out land): The wind has removed most of the soil profile and the land is classified as a miscellaneous land type. Use of the land for ordinary agriculture is not feasible without extensive reclamation. Blown holes are numerous and deeply carved into the lower soil parent material.

Classes of Areas with Significant Material Deposition (Soil Survey Staff[50])

Class 1 (Overblown): Recent deposits of wind-drifted material cover the soil in layers thick enough to alter its characteristics significantly up to 24 inches.

Class 2 (Wind hummocky): Recent deposits of wind-drifted soil material in a fine pattern of hummocks or low dunes.

DESERTIFICATION, A GLOBAL CHALLENGE

The problem of desertification has become worldwide. Its disastrous effects threaten man's efforts to produce more food for the fast-increasing world population. As pointed out in the preceding pages desertification is more severe in arid ecosystems. Efforts to combat desertification are not always capable, for one reason or another, to successfully combat this problem. Soil desertification has been taking place in different regions of the globe due to its various causes (Table 45). The result of such deterioration has always been reduction of land productivity or even complete failure of crop growth.

National and international efforts have cooperated since the Sahel catastrophe to save the areas vulnerable to desertification and cure the desertified soils.

Though actual success is slow, understanding of the size of the problem has become much clearer than a decade before.

However, further efforts, investments, and persistence are still needed.

International Efforts

Though it is not intended to give a full account of various international organizations' activities to combat desertification, the following pages may give examples of these activities.

Efforts of the international organizations to disseminate knowledge, and inform governments and the people with facts concerning the severity and catastrophic effects of the desertification problem are variable. They have held conferences and symposia, published books and magazines, prepared projects to combat desertification in vulnerable regions, and raised funds to execute such projects in countries which are not, for one reason or another, able to achieve these projects.

In December 1974 the United Nations General Assembly passed a resolution 3337 (XXIX) (UNDESCON Secretariat[14]), calling for an International Conference on desertification to be held in 1977. The General Assembly specified (1) to prepare for this conference a world map developed for the areas affected or likely to be affected by desertification, (2) all information on desertification and its consequence for development should be gathered and assessed, and (3) a plan of action to combat desertification should be prepared with emphasis on the development of indigenous science and technology.

The UN Conference on Desertification[8] (UNCOD) was convened in Nairobi, Kenya, 29 August to 9 September 1977 and recommended a Plan of Action to Combat Desertification which was endorsed by the General Assembly, including the following recommendations.

"Rec. 6: That measures should be taken to prevent desertification and ameliorate the condition of degraded rangelands, to introduce suitable systems of rangeland, livestock and wildlife management to develop diversified and integrated systems of production and to improve the living conditions of the inhabitants of these areas."

"Rec. 7: That comprehensive measures should be adopted for the conservation and improvement of soil, and rational use of soil moisture in order to prevent and fight desertification in rainfed agricultural areas."

Table 45. World area affected or likely to be affected by desertification (000 km²).

Region	Existing Extreme Desert	Degree of Desertification			Total Land Area Affected	
		Very High	High	Moderate	000 km²	%
Africa	6178	1725	4911	3741	1655	55
North & Central America	33	163	1313	2854	4363	19
South America	200	414	1261	1602	3478	20
Asia	1581	790	7253	5608	15232	34
Australia	—	308	1722	3712	5742	75
Europe	—	49	—	190	238	2
Total	7992	3449	16460	17707	45608	35

Source: FAO.[97]

"Rec. 8: That urgent measures be taken to combat desertification in irrigated lands by preventing and controlling water-logging, salinization and alkalization by reclaiming deteriorated lands, by improving irrigation and drainage systems, by modifying farming techniques to improve productivity in a regular and sustained way, by developing new drainage and irrigation schemes where appropriate, always using an integrated approach, and through improvement of the social and economic conditions of people dependent upon irrigation agriculture."

"Rec. 9: That existing vegetation be maintained and protected and special measures be taken to revegetate denuded areas and then maintain and protect them to promote soil conservation, and to stabilize moving sands. This may be necessary in areas where human activity may have an adverse environmental impact on hilly areas and on mountain slopes, particularly at points where deterioration may threaten settlements, roads, farmlands and reservoirs and along vulnerable desert margins."

In order to implement these recommendations, it is not only necessary to obtain a comprehensive assessment of each of the desertification aspects, factors involved, and its impact on people of the desertified area, but also it is necessary to obtain an assessment of the requirements, including financial resources of implementing programs to combat desertification and estimate of the share that international funding may be requested to provide. Accordingly, the costs of programs based on as factual estimates as possible represent a basic step in such an international venture (Balba[22]).

Any plan to combat desertification depends on knowing the size and type of the problem, hence monitoring becomes a basic step preceding planning and implementation. National and global systems of monitoring are thus required.

World Soil Charter (FAO[68])

In the 21st session of the FAO in Rome, November 1981, decision No. 8181 was issued proclaiming the "World Soil Charter."

The following is an outline for the charter.

Referring to the Decision No. 6 of the World Food Conference (Rome, 1975) which urged FAO to approve World Soil Charter as a basis for world cooperation for the benefits of the world soil resources, and as it became certain that these soil resources are limited and that only a small proportion of the total world soils is utilized for food production for the world inhabitants expected to be about 6 billion at the end of this century, and referring to the Work Program adopted by the World Conference for Land Reform and Rural Development (Rome, 1979) which called for "the efficient utilization of land with careful attention for ecological equilibrium" and UN Work Program to Combat Desertification (Nairobi, 1977), and being convinced that the need for feeding world population can be satisfied including malnutrition.

The Conference recommends:

1. Increasing the food production including intensive agriculture wherever possible.
2. Utilizing the uncultivated area wherever sustainable agriculture is feasible.
3. The correct utilization of pasture land and forests. After expressing the fear from soil deterioration resulting from misuse of the land and the faulty actions taken to improve the production especially in regions subject to water or wind erosion or salinization. The conference endorses the "World Soil Charter" and urges the United Nations and the international organization concerned to support through their channels and activity fields the following principles and plans:

Principles

1. The land is one of the most important resources for man because of the water it contains and plants and animals related to it. Utilization of these sources should not cause their deterioration or desertification. The existence of human beings is tied with the sustainable productivity of land.
2. With the recognition of the importance of the soil resources for the existence of man and his well-being and the countries economical independence as well as the increasing need for more food production, it is necessary to give the first priority to the soil, its conservation and improving its productivity.
3. By soil deterioration it is meant its total or partial loss of its productivity either in quantity or quality or both as a result of various processes such as water or wind erosion or increase in its salinity or sodicity or waterlogging or the depletion of its fertility or losing its vegetative cover. In addition, there are vast areas lost daily due to the nonagricultural utilization. These processes raise concern in the light of the increasing need for food, fibers, timber, and production.
4. Soil deterioration directly affects the agriculture and forests and decreases their production and disturbing the water course as well as other sections in economics and environment in general including industry and commerce due to floods, and silting of rivers, reservoirs, and harbors.
5. Governments are responsible for including measures that guarantee the safe utilization, permanent conservation, improved productivity, and preservation of productive land in their programs for land utilization. Participation of land users is necessary to ensure sound land utilization.
6. A participation of the system of incentives at the village level as a part of legal frame to insure the efficient utilization of the soil should be initiated.
7. The relief or allowances which may be offered to the farmers should aim to encourage them to care and apply the soil conservation measures.
8. The land owners-tenants relationships may be an obstacle in some cases for adopting the sound soil management and conservation at the farm level. Accord-

ingly, these obstacles should be eliminated especially those related to rights and obligations of the owners and the tenants in the light of the recommendations of the World Conference for Land Reform and Rural Development (Rome, 1979).

9. Land users and the public should be informed about the importance of improving the land productivity and its conservation and the means to achieve it, giving much care to education and extension programs and training the responsible officials of various levels on carrying out these programs.

10. Each country should evaluate the soil suitability—at every input level—for various utilizations: rain-fed or irrigated agriculture or forestry to insure the maximum benefit from the land.

11. Utilization of the land which suits multi-utilization should be elastic to allow in the future other utilizations. Also, utilization of the soil in nonagricultural purposes should be planted to avoid permanent deterioration or occupancy of the soil.

12. Any decision concerning soil utilization and management should observe the permanent benefit of the soil and not short period gains which may lead to ruining the soil resources.

13. Soil conservation measures should be included in the soil development plans and their expenses.

Manipulation

The Governments

1. Prepare a plan of the soil utilization according to its suitability for various utilizations and the country needs.

2. Formulate the principles of soil utilization, management, and conservation in the suitable legislative forms.

3. Impose a system for soil monitoring, supervising, managing, and conserving in the legislative form to be coordinated with other governmental organizations concerned with soil resources.

4. Evaluate the suitability of the new land and the lands presently under cultivation, for variable utilizations and possibilities of being deteriorated, suggesting other options which may better-suit the soil capability.

5. Conducting education, training, and extension programs at variable levels in soil management and conservation.

6. Disseminate information related to soil erosion and measures to combat it in the farm and in watersheds emphasizing the importance of the soil resources for the development and for the people.

7. Strengthen the relationships between the governmental agencies and the land users in order to manipulate the soils policy. Attention should be made to carry out the technologies of soil conservation and the integration between protection process of forests and agriculture and environment protection.

8. Attempts to create social, economical legislature conditions to suit the efficient management of soil resources. These conditions should include a system for land lease incentives for the land users such as subsidies, decreasing taxes, and facilitating loans. Encourage the universities which wish to cooperate with each other and with the government to achieve the efficient use of soils, their conservation and improvement.

9. Conduct field research which provides sound scientific techniques for soil improvement and conservation which take into account the social values and the economic conditions.

The International Organizations

1. Intensify efforts to create understanding and encourage cooperation among all international sectors by offering the help where it is needed and assisting the information media and call for meetings at various levels and facilitate acquiring the technical publications.
2. Assist the countries especially the developing, upon their request, in formulating the legislative systems and the means which enable them in conducting, monitoring, and following up the sound practices of soil and its conservation programs.
3. Support cooperation among the governments which adopt the sound practices of soil especially in rainfall regions.
4. Pay special attention for the land development requirements including conservation and improvement of soil resources, providing the production requirements and incentives at the farm level and rainfall regions, and establishing the necessary organizations.
5. Strengthen programs related to soil conservation, not only those of technical nature but also research in social fields which are related to conservation and management of soil resources.
6. Collect, store, and publish experiences and information related to soil conservation programs and the results obtained in various environmental regions.

World Soils Policy[70]* (Garbouchev and Ahn[71])

By decision 7/6 B, 1979, the Governing Council of UNEP requested the convening of a high-level group of experts to identify the most important scientific, technical, constitutional, and legal elements of soils policy.

The initial meeting was held in Rome in March 1980 with a subsequent meeting in 1981, again in Rome, which was convened in cooperation with FAO and UNESCO.** By decision of Governing Council 8/10, 1980, the Executive Director of UNEP was asked to work in collaboration with relevant UN and other international agencies to prepare for adoption of a world soils policy, to develop a Plan of Action for implementing the policy, and to encourage and support development and introduction of soils policies at national levels.

The UNEP decision to foster development of world and national soils policies and formulation of a Plan of Action was made with the knowledge that:

a) despite short-term improvements due to technology, the world's long-term capacity to grow food and fibre is being reduced by continuing loss and degradation of its soils, and
b) that in many countries, there is a lack of recognition of the problems on a national level and a lack of resources to attack and overcome these problems.

A world soils policy forms an integral part of the World Conservation Strategy which has as its main aim to conserve the living resources for sustained development. A number

---

*   Expert Meeting on Plan of Action for World Soils Policy, Part 2. (This book's author was a member of the group of experts.)

** A third meeting was held in Geneva in 1982.

of the priority issues identified in the World Conservation Strategy are those directly concerned with maintaining soil productivity, prevention of soil degradation, and stimulation of soil reclamation.

The need for the development and adoption of world and national soils policies is further accentuated by the rapidly increasing world population and the steadily growing "lifestyle expectations" of the world's people.

There are many causes of soil degradation, ranging from those associated with agricultural and pastoral use to those resulting from mining and other "non-rural" uses. Similarly, in the past, there have been many effective soil improvement and protection programs undertaken by national and international bodies.

Objectives of a World Soils Policy

A world soils policy Plan of Action seeks to:

- increase and apply scientific knowledge of the soils of the world to their potential for production, and their sound management,
- encourage and assist countries in improving the productivity and management of their soils and in reducing soil degradation,
- encourage the better management and conservation of soil resources, reduce pollution, and improve the quality of water and air,
- develop and promote agricultural production systems which assure the use of the soil on a sustained basis,
- enlarge and improve the world's supply of arable agricultural land through irrigation, flood control, and reclamation,
- slow the loss of productive agricultural, pastoral, and forest land to other purposes,
- monitor changes in soil quantity and quality and in land use, and
- bring to the attention of the people of the world, and their political leaders in particular, the extent of world soil degradation and its seriousness, its causes and its remedies, and is addressed to international and national bodies, as well as individuals with interests and responsibilities for safeguarding soil, water, and related resources.

Orientation of the Plan of Action

In developing the Plan of Action for implementation of the world soils policy at national and international levels there has been emphasis on identification of vulnerable areas and on the needs of developing countries in particular, as well as on research and education and in bilateral, national, regional, and international transfer of knowledge and experience.

International action required to promote sound use of land and soil resources has been identified and the need for internationally acceptable techniques for recognition and for prevention of soil degradation and achievement of reclamation has been highlighted. The role of international bodies lies particularly in the field of establishing broad priorities in the field of achieving resource knowledge and experience transfer, as well as in supervision of internationally funded programs, and in the initial stages should assist individual national governments in formulation of their own national soils policies.

At the national level, government commitment to sound use of soil resources is required, along with associated legislation, educational, and research inputs, if action to control soil degradation is to be successful. Once a national policy has been adopted, it should act as a guide in all aspects of national development.

UNEP and the Mediterranean Action Plan (MAP)

United Nations Environment Programme (UNEP) was established in 1972 by the decision of the UN General Assembly as a specialized UN agency. The task committed to UNEP was to act coordinatively and catalytically in environment protection activities by means of adequate programs.

The Mediterranean Action Plan (MAP) is one of the UNEP's Regional Seas Programs initiated in 1975.

In 1985, the Regional Activity Center of the Mediterranean Action Plan,[72] UNEP, called for a meeting of the experts representing the countries of the Mediterranean coast to discuss a project entitled "Promotion of Soil Protection as the Essential Component of the Environmental Protection of Mediterranean Coast Zones" (Project No. PA-8/M-E/510283-0572). In that meeting, national reports were presented describing several aspects of land use and degradation in each country. The climatic conditions of the southern coast of the Mediterranean are somewhat different from those of the northern coast. In the south, the temperature is higher and rainfall is generally less than in the north coast.

Each country especially in arid and semiarid regions has become aware of the desertification problems and tried independently and/or assisted by the international organizations to carry out projects to protect its land and cure the areas which were desertified.

In 1978, the FAO Regional Office for the Near East in Cooperation with Soil Resources, Management and Conservation Service, Land and Water Development Division of the FAO, Rome, assigned Rafiq[73] to survey the extent of the desertification problem in 10 countries in the Near East Region.

The following activities were among the consultant's responsibilities:

- To assess actual soil degradation and soil degradation hazards.
- The assessment is to be based primarily on existing data.
- Interpret the environmental factors which influence the extent and intensity of soil degradation.

Rafiq's report included evaluation of soil degradation in Egypt, Iran, Iraq, Jordan, Sudan, Somalia, Libya, and Tunisia. The assessment of the degree of soil degradation presented in Rafiq's report is more on the qualitative side than that of the FAO/UNEP/UNESCO presented above.

Sant'Anna[74] summarized the recommended plan of action in the FAO Regional Conference of 1986 at Yamoussoukro, Cote d'ivoire, which was approved by the ministries of agriculture attending the conference. As a response to this recommendation, the Land and Water Development Division of FAO in cooperation with African experts produced two volumes:

a. The International Action Program on Water and Sustainable Agricultural Development, and
b. The Conservation and Rehabilitation of African Lands, an International Scheme.

The latter publication emphasized the importance of the rural people's participation in finding and applying improvements to existing systems which would be visibly beneficial in providing food and financial security.

The framework for the plan is:

At the national level, the following action can be followed in any country in developing effective conservation and rehabilitation programs of its resources:

1. The first involves improving land use by evaluating land resources and establishing a land resource data base.
2. To implement the land improvement, a series of parallel routes may be taken:

   - Reform agricultural price strategies,
   - Introduce appropriate technologies,
   - Make farm inputs more easily available,
   - Reform land tenure system,
   - Diversify rural income, and
   - Relocate land users.

3. Develop national institutions. It may be necessary to create a high-level advisory committee to formulate conservation strategy and development policy, coordination, and monitoring.

According to Kassas,[75] only modest progress was achieved since 1977 after the UN Conference on Desertification. In 1992, the UN conference on Environment and Development resolved to establish an "Intergovernmental Negotiation Committee for Elaboration of an International Convention to Combat Desertification." The Committee has held a series of sessions and plans to produce a convention for signature in 1994. Kassas concluded that it is hoped that "this will provide the world with an effective tool for managing the desertification menace."

National Efforts to Combat Desertification

The arid and semiarid regions of the world cover about one third of the earth's surface. Large parts of these regions became desertified during past centuries, but the process is still continuing and many vulnerable areas are turning into desert.

The problem of desertification has been highlighted by the disaster of the Sahel and parts of Asia and Latin America. At present, desertification threatens the livelihood of some 50 to 90 M people living in the drier areas of the world (UNESCO[76]).

Desertification is a plight that destructs the productive land and causes miseries, poverty, and hunger to its benefactors. In 1980, UNEP cited by Goudie[2] estimated that desertification has occurred in 61% of 3257 M ha of productive drylands. From this area, 2556 M ha is rangeland, 570 M ha rain-fed cropland, and 131 M ha irrigated land. Desertified land percentages in these areas were estimated to be 62, 60, and 30, respectively. FAO[38] estimated that desertification threatens 35% of the globe's land surface and 19% of its population.

In North America, three instances of widespread desertification have occurred during the past 200 years: overgrazing in the desert grasslands, salinization of irrigated land, and wind erosion in the Great Plains in the 1930s (Dregne,[9] Table 46).

Table 46.   Arid lands affected by desertification in North America.

Country	Irrig. Land (000 ha)		Rain-Ted Cropland (000 ha)		Rangeland (000 ha)	
	Total	Affected	Total	Affected	Total	Affected
Canada	300	60	5000	3000	10000	7000
Mexico	3750	1125	7500	6700	100000	96000
USA	15500	1650	30000	15000	235000	188000
Total	19550	2835	42500	24700	345000	291000

Source: Dregne.[9]

Deterioration of the vegetative cover in the cool arid regions of the northern United States and of Canada was seldom as severe as it was in the hot arid regions of the southern United States and northern Mexico. Four grazing areas in the southwestern United States have experienced very severe desertification that has essentially ruined the land for the foreseeable future (Dregne[9]).

Dreen[77] described the "Dust Bowl" disaster as follows: unlimited number of farms were in a very bad condition. Everywhere heaps of sand covered and filled backyards, buildings' roofs, machines, water reservoirs, and canals. The fields appeared striped with dry nuded nonproductive sand. The dust storms prevented the sun light and it was as dark by day as it is at night. Livestock died from hunger or because of the dust they swallowed. Machines and cars were destroyed and roads become unlevel. Wind erosion was so much that dust reached the eastern part of the United States and 3000 km in the Atlantic Ocean.

In the second half of the 19th century, the southwestern region of the United States attracted the activities of the settlers with its wide pasture land. Because the new inhabitants did not apply the traditional grazing practices which conserve the plants and the soil, unpalatable plants started to invade the pasture and to replace the grass. Thus, the pasture regions started to deteriorate as a result of overgrazing by excessive numbers of livestock.

From 1910 to 1920, people changed the unsuitable grazing practices to save their pasture land. They also controlled the number of livestock. But after 20 years the situation did not improve. This was due to introducing irrigation into the region. Well waters were used to irrigate saline or impermeable soils occupying low-lying areas. The soil water table rose causing the accumulation of salts on the soil surface.

The soil suffering from salinization spread out in southern California. With the increase in salinity, yields decreased and farming salt-affected soils became uneconomical. The solution to the salinity problem was the establishment of efficient drainage systems (Dreen[77]).

The arid lands of South America are found mostly in the narrow southern part of the continent and in northeastern Brazil. Soil degradation is mainly due to water erosion first, followed by wind erosion and salinization, both of which are extensive in Argentina (Table 47, Dregne[9]).

Kilinc[78] stated that causes of wind erosion in Turkey are misuse of land, overgrazing, forest fires, misuse of vegetation, rugged topography, and sensitive parent materials.

Wind causes the sand dune movement toward the inland during the dry season. Intensity of erosion is characterized by the removal degree of the "A" horizon of the soil profiles.

In Syria, Nahal and Darmash[52] stated that water erosion may exceed 200 t per ha per year, especially in the mountainous coastal region and the Kurd high mountains region. In Homs, soil erosion by water is about 50 to 200 t per ha per year. In other regions it is

Table 47.   Arid lands affected by desertification in South America.

Country	Irrig. Land (000 ha) Total	Irrig. Land (000 ha) Affected	Rain-Fed Cropland (000 ha) Total	Rain-Fed Cropland (000 ha) Affected	Rangeland (000 ha) Total	Rangeland (000 ha) Affected
Argentina	1550	310	5000	3800	180000	126000
Bolivia	65	6	1000	950	12000	11500
Brazil	520	78	6000	5000	140000	135000
Chile	1280	320	1400	1350	24000	22400
Colombia	0	0	0	0	3500	3200
Ecuador	460	115	40	39	300	280
Paraguay	9	2	50	20	12000	9600
Peru	1155	346	500	450	9500	8800
Venezuela	350	52	300	250	2800	2600
Total	5389	1229	14290	11859	384100	319380

Source: Dregne.[9]

about 10 to 50 t per ha per year. These figures are considered high as these areas have been eroded in past years.

They also stated the vegetation cover in Kalamon, the Eastern Lebanon, and the Badia is severely deteriorated.

Rafiq's[73] report stated that cultivating areas having annual rainfall less than 250 mm to wheat and barley required clearance of natural vegetation. Because cropping takes place usually every other year, the unprotected soil is subject to wind erosion. The government prohibited by law cultivation in areas having less than 200 mm of annual rain. A considerable proportion of the rangeland abides with the "Range Management Program" of national grazing and supplemented feed.

Waterlogging and salinity are problems in the irrigated areas of the Euphrates Valley (in Syria). Drainage networks have been installed. Gypsum-rich soils are frequently encountered.

Rangelands in Jordan are defined as areas receiving less that 200 mm of annual rainfall. This area is about 8.5 M ha or about 90% of the total area of East Jordan.

The rangeland in Jordan is deteriorating at an alarming rate due to overgrazing, fuel wood-cutting, and unsound cultivation combined with periodic droughts. The sparse and scattered vegetation in the dry area which receives less than 100 mm of annual rainfall consists mainly of few *Artesemia herba alba*, *Anabasis* spp., *Retama ratatam*, *Zilla spinosa*, *Aristida* spp., *Calliaonum commosum*, *Salvia dominica*, and *Ballota undelata*, most of which are unpalatable species.

Areas receiving 100–250 mm of annual rainfall are considered the best rangeland in Jordan and contain palatable nutritive grasses and shrubs. However, most of this area has been subject to unsound cultivation for cereal production, uprooting of shrubs for fire wood, and early grazing which resulted in complete destruction and elimination of palatable species (Balba[79]).

Areas receiving more than 200 mm annual rainfall are subject to erosion. These areas include the Western highlands. The hills are so eroded that the rocks are exposed.

Efforts to control water erosion are practiced on sloping lands and hills by afforestation using olives, grapes, and apple trees. Wind erosion is a serious problem in the desert and semidesert (Balba[79]).

In the Eastern Province of Saudi Arabia, rapid economic development has been initiated as a result of the oil fields and industrial activities.

This region within the Jafurah and sea is known with its strong unidirectional winds, abundant sand dunes, and less than 80 mm of annual rainfall. The sand control problems are immense (Kerr and Nigra[32]). About 10 M trees were planted by 1985. Still more trees are needed.

Agricultural activities in Saudi Arabia are variable. Water control, constructing dams, lined canals, digging wells, and variable methods of irrigation are considerably increasing the agricultural production in the Kingdom.

Rafiq[73] stated that six types of soil degradation are important in Iraq, namely water erosion, wind erosion, salinity, waterlogging, flooding, and depletion of soil fertility.

Water erosion is an important problem in about 5 M ha. The problem was increased by extending crop production in regions receiving 200 to 300 mm annual rainfall. A program of watershed management was started. Other water and soil conservation measures need to be applied.

Wind erosion is important in the desert and semidesert areas. Also, wind erosion is active in some parts of the lower Mesopotamian plain where dunes are formed by accumulation of granules of saline soil material.

In the lower Mesopotamian plain salinity is the major problem. The development of the secondary salinity has spread out. It is thought that the physiographic position of the area in the flood plain influences the hydrology and the overall salt balance of the area. Waterlogging is associated with salinity in about 25% of the lower Mesopotamian plain. New irrigation projects are equipped with drainage systems.

The cultivated area in Oman Sultanate is 83360 ha while the rangeland occupies about 500,000 ha. The deterioration of rangeland increases due to overgrazing and wood-cutting. Sand encroachment also destroys the natural vegetative cover especially in Sharkiyah Region which is adjacent to Wahebah Desert. A pilot project was suggested to protect the Al Wassil Area in Alsharkiyah Region from blown sand and to improve the vegetative cover (Balba[80]).

In the USSR, the process of water erosion takes place in about 15 M km^2. Also, the insufficient moisture and overdrying of the plowed areas inevitably lead to a high degree of susceptibility to wind erosion. According to Gerasemov[81] wind erosion is not less dangerous than water erosion.

In Iran, wind erosion and sand encroachment are problems in the desert and semidesert regions. About 18 M ha are covered by sand in Khuzistan. Micro-windbreaks 1 m high were established in rectangles 10 × 75 m. Other patterns were also tried and compared. It was found that windbreaks 7 m apart and 0.5 m high were as efficient as other windbreaks and at the same time cost less than the others. Studies showed that sowing seeds was not satisfactory in fixing the sand area, neither did the Tamarix spp. (Bhimaya[30]; Niknam and Ahranjani[82]).

Water erosion is active even in the gently sloping land under cultivation in the high plateau of Azerbaidjan and Kurdistan (Iran) where annual rainfall is more than 300 mm.

Primary salinity prevails in the deserts along the southern coast and between Ahwaz and Abadan. Secondary salinity is due to irrigation without drainage. The problem was controlled by a drainage system (Rafiq[73]).

In Pakistan, human activities have accelerated the process of soil erosion during the last 100 years or so by overgrazing and overcutting of vegetation in the cultivated areas and expansion of cultivation on very sloping lands and by adopting crop rotations that are not suitable for controlling erosion. Classification of areas according to the erosion hazards is given in the following (Ministry of Food and Agric., Gov. of Pakistan[83]):

- cultivated area with minor erosion problem 3040 ha,
- cultivated area with moderate erosion problem 242240 ha,
- cultivated area with severe erosion problem 38200 ha.

Rafiq's[73] report stated that salt-affected soils cover 7.2 M ha in Pakistan out of which secondary salinity covers about 3 M ha. Water erosion causes much damage to about 2.7 M ha distributed in several regions. The soils in the Tal Area, Bannu Basin, and the Southern Desert suffer severe wind erosion.

Various activities are taking place to survey the land, define the problems, and apply the necessary practices which suit each situation.

According to Mann,[84] the sown area in the arid zone of Rajastan, India, increased from 28.6% in 1951 to 54% by 1971. A part of this increase was at the expense of pasture and rangelands which decreased by about 25%. Accordingly, the fodder supply was reduced and soil erosion was enhanced. Although the crop area has increased, the production per unit area has considerably decreased, indicating that the fertility of the land has declined.

Studies have shown that due to human interference, sand movement caused land undulation (CAZRI[85]). Recent sand activity has led to a 15 to 30 cm increase in sand thickness on fence lines and widened them 1 to 2 m. Studies also showed that 4.25% of western Rajastan has already been desertified, 76.15% of the area has been categorized as highly to moderately vulnerable, and 19.5% as moderately to slightly vulnerable to various processes of desertification. A study by Gupta and Aggarwal[86] as stated by Mann[84] showed that from a bare sandy plain, there was a continuous increase in the removal of sand. Incoming water into desert land may affect the ecosystem adversely. The water table rose in several areas in the irrigated region at an average rate of 1.52 m per year.

To combat desertification, successful technologies were demonstrated in about 60 centers spread over the entire western area of Rajastan. These technologies aimed at minimizing soil erosion from the bare dunes, covering the denuded soil surface by plant species, minimizing the adverse effects of strong hot winds, using shelter belt plantations, and using the optimum amount of irrigation water (Table 48, Tejwani[87]).

According to Walls,[88] the arid regions of China are mostly typical of the temperate zone arid lands. Most of these lands receive less than 200 mm annual rainfall and with annual evaporation reaching as high as 3500 to 4000 mm. Shifting dunes occupy about 75% of the sandy soils. The Taklimakan Desert in Xinjiang with an area of 327000 km^2 is the largest desert in the country. It is famous for its moving large sand dunes 100–150 m high.

Protecting the oases and the cultivated lands from blowing sand is the major activity in combating desertification. Farmland shelterbelts are preferably composed of fast-growing indigenous plus a small proportion of long life species. The following characteristics are important in selecting the plants used to stabilize the sand dunes:

- Resistance to heat and cold.
- Resistance to drought.
- Long growing period.
- High nutritive value.
- Adaptability.
- Long life.
- Easy to work with.

The studies of the Department of Desert Research, Academica Sinica, introduced *Caraaona* spp. to satisfy the above stated characteristics.

Table 48.   Extent of soil conservation problems in India.

Type of Land	Area (million ha)	
	Total	Soil Cons. Prob.
Forest	61.2	20
Cultivated waste lands	17.4	15
Permanent pasture and other grazing lands	14.8	14
Land under miscellaneous tree crops and groves	4.2	1
Fallow land		
Fallow land other than current fallow	9.2	8
Current fallow	11.1	7
Net area under cultivation	137.9	80
Land not available for agriculture or forestry	50.2	—
Total	306.0	145

Source: Tejwani.[87]

Because most of the moving sands are located in semiarid zones, favorable water and energy conditions offer a good chance of combating sand movement by vegetative means (Walls[88]).

About 55% of the African land area consists of existing deserts or is subject to desertification of various degrees (Table 49, FAO[38]).

According to Tolba,[20] "On the southern fringes of the African Sahara 650,000 km^2 (65 M ha) of once productive land have become desert during the last 50 years; about 60,000 km^2 (6 M ha) of productive and fertile land are lost annually; some 600–700 M people (14% of the world population) live on threatened drylands: of these about 60 M are immediately affected by desertification" (Dr. M.K. Tolba, executive Director of UNEP, 1978).

According to Ayoub,[89] 74% of rangeland and 61% of rain-fed cropland in this continent are affected by desertification. The productivity of the savanna, where most of Africa's population lives, has dropped by 35%. Over 50 M people live on land so badly degraded they are frequently at risk from famine. He added that a recent (1990) UNEP/ISRIC publication shows that about 49 M ha of African land surface are affected by degradation, 46% by water erosion, 38% by wind erosion, 12% by chemical deterioration (mostly loss of nutrients and salinization), and about 4% by physical deterioration, largely by compaction. About 124 M ha of the African soils are classified as strongly degraded and 192 M ha are moderately degraded (FAO[38]).

Rangelands in Africa

The rangeland in Africa occupies several geographic and climatic regions (Balba[43]):

- The Mediterranean basin where rainfall is concentrated in the relatively cool winter season, thus permitting reliable cropping extending to areas of 300–400 mm annual rainfall.

Table 49.   Irrigated and rain-fed area in Africa (million ha).

	1980	2000
Irrigated	4	6
Rain-fed	200	250
Total	204	256

Source: FAO.[57]

- The Sahelian belt, fringing the Sahara, characterized by a very hot dry season of 7 to 10 months, with rainfall concentrated between May and September.
- The east African equatorial belt, influenced by the presence of highland masses, where much of the range areas enjoy bimodal rainfall and warm but relatively equable temperatures.
- South Africa where rainfall is concentrated mainly between September and March, with a relatively cool dry season.

Within each of these climatic regions, there is considerable variation and land potential, depending on the site conditions.

There is great variation in the pastoral systems between and within these four regions. The disaster of the Sahelian region is described in the following pages.

In the Nile Delta and Valley, the main soil problem in Egypt is secondary salinity. However, after the installation of an elaborate network of open and tile drainage system, this problem became under control.

In the far northern part of the Delta, the origin of salinity is the inundation by saline lakes' water.

Leaching these soils was carried out by using drainage water. Accordingly, its complete reclamation had to wait for the fresh Nile water after the construction of the High Dam at Asswan.

The newly reclaimed lands at the fringe of the Western Desert added to the cultivated land also suffered from waterlogging and secondary salinity because the drainage system of this region had to wait 10 years before its completion.

The Government established the Authority of Land Improvement to carry out deep plowing, application of gypsum, and improvement of the drainage system in small holdings.

Water erosion, sand dunes creep, and wind-blown sand usually take place in the North Coast. Wind erosion control was practiced by establishing windbreaks in the newly cultivated areas along the desertic parts of the Mediterranean coast. Stabilizing sand dunes was carried out at an experimental scale. Also, water conservation practices had recently started in several wadies in the northwest coast region. Wind erosivity index in the northwest coast is 150 t/ha/yr, in the middle part of the coast is 50–100 t/ha/yr, and in the eastern part (Sinai) is 50–100. The rainfall erosivity index is 50 t/ha/yr which is considered moderate (Balba[44]).

The El Sahel region extends from west of Africa eastward separating the Sudanese agricultural land from the Great Sahara. Its borders depend on rainfall. They widen in drought years and shrink in the rainy years. The region occupies a considerable area of about 4 M km² of nearly barren land extending from Mauritania and Senegal on the Atlantic Ocean eastward crossing Maly, Upper Volta, the Niger to Chad, and the Sudanese borders (Ayoub[89]).

In the portion bordering the Sahara, annual precipitation is 100–350 mm and increases to 350–600 mm in the southern portion adjacent to Sudan and Chad where rangemen and farmers live together and exchange meat and hide for wheat and barley. Though rainfall may reach as much as 600 mm, it falls in short vigorous torrents and the rate of evaporation is so high that about 80% of rain is lost by evaporation. The region used to suffer from drought, the last of which took place in 1940. In 1968, drought returned. Rainfall dropped from 284 mm to 122–149 mm and to the low of 72 mm in 1972.

In the period 1968 to 1975, the Sahara extended about 150 km to the south. In the fifth year of drought, Lake Chad shrunk to one third its normal area. In the preceding winter, the Niger and Senegal rivers had failed to flood leaving much of the cropland barren. Pasture land in the north of the Sahel was depleted causing a widespread death of shrubs and trees.

Weakened livestock concentrated around larger watering points where vegetation and soil were destroyed.

Estimates of the people who died as a result of this disaster have ranged between 100,000 and 250,000 (UNCOD Secretariat[8]).

To explain the disastrous situation of the last drought, which had different causes than the several droughts in the past, Brine[90] pointed out that several changes have been introduced in the region to improve the living conditions and alleviate the miseries of the people, such as digging deep wells, building water basins for the livestock, and controlling their epidemics.

Also, crops such as cotton, peanuts, and rice were introduced in the agricultural areas. When the inhabitants' number increased they farmed the neighboring pasture land.

With combating livestock diseases their number increased and more pasture land is needed, while this land had shrunk after being cultivated, partially, to field crops. These herds crowded close to water sources and damaged the pasture plants around them. Improving the drinking water problem caused a problem of hunger. It is estimated that one third of the livestock died from hunger.

For the Sahel Region from Senegal to Sudan a pilot transnational green belt was justified by reason of the homogeneity of the ecological conditions throughout the region.

Rehabilitation and management activities should deal mainly with restoration and efficient utilization of the pastoral vegetation, both herbaceous and leguminous, the improvement of herd management, and the organization of livestock owners. To implement the project, it was suggested that each country involved set up a national committee responsible for demarcating priority action units (UNEP[17]).

Libya's activities to stabilize sand dunes are worth mentioning. An extension bulletin published by the Libyan Ministry of Agriculture[31] and a report by Suleman described these efforts in the Mediterranean Action Plan Meetings as follows:

1. Utilization of fences surrounding the farms made of plants such as the cactaceae.
2. Mechanical means.
3. Windbreaks.
4. Utilization of plant residues in tunnels 15 cm deep with about 35 cm above ground in rectangular formation.
5. Utilization of asphalt.

The heated batwoman to 50°C is spread under pressure of about 2.7 kg per cm^2 at the rate of 2.5 ton per ha in level areas or 1.6 kg/ha in hills. Trees are planted before spraying the batwoman.

The seedlings 6–10 months old are grown in pots until they are 60–80 cm long. When the soil is relatively moist after about 40 mm of rainfall and the moist depth is about 40 cm, the seedlings are dipped in water before planting. Half of the tree seedling length should be above the ground level. The following trees proved successful in continental

Table 50.   Percentage of land requiring treatment for various forms of degradation in the arid zone of Australia, June 1975.

Type of Degradation	Extent
Vegetation degradation and little erosion	51
Vegetation degradation and some erosion	25
Vegetation degradation and substantial erosion	15
Vegetation degradation and severe erosion	8
Dryland salinity, sometime in combination with water erosion	<u>0.06</u>
Total treatment needs %	100

Source: Logan.[93]

dunes: *Acacia coamophylla, Eucalyptus camaldulensis,* and *Eucalyptus gomphocephal.* In coastal sand dunes, *Acacia cocloris* and *Acacia cyanophylla* were more successful. Rubber emulsion is also used experimentally.

Water erosion in the forms of sheet and gully erosion occurs in the coastal zone of Lybia. Also, large areas in the region are affected by blown sands. Sand dunes fixation receive considerable attention as stated above (Suleman[72]).

The same problem of water and wind erosion in Tunisia was emphasized by Souissi.[91] Efforts to protect the soil included afforestation of 15,000 ha of forest belts. Trabolsi stated in Mediterranean Action Plan Meetings[72] that erosion has taken place in the coastal zone due to devastation of forests and overgrazing. Erosion was extended to about 30% of the total area. The intensity of soil erosion is about 16 and 30 m^3/ha/yr in humid and semiarid eastern regions, respectively. Efforts to control soil erosion by water and by wind included afforestation of vast areas (250,000 ha) and soil protection in 51,000 ha. In Morocco, erosion protective action was intensified in 1971. Several projects to control water erosion were carried out in cooperation with FAO as NEKOR and DERRO projects in 78,000 and 25,000 ha, respectively (Abdel-Kader[92]). In Australia, Logan[93] estimated the percentage of land requiring treatment for various forms of degradation in the arid region as presented in Table 50.

REFERENCES

1. Kropotkin, P., The desiccation of Eur.-Asia. Geographic J., 23:732, 1904, cited in Goudie,[3] 1990, p. 3.
2. Goudie, A.S. (Ed.), Techniques For Desert Reclamation. John Wiley & Sons, Chapter 1, 1990.
3. Jones, B., Desiccation and the West African colonies, Geographical Journal, 91:401, 1938, Cited in Goudie,[2] 1990.
4. Stump, I.D., The southern margin of the Sahara; comments on some recent studies on the question of desiccation in West Africa. Geog. Rev., 30:297, 1940. Cited in Goudie,[2] 1990.
5. Aubreville, A., Climats, Forets et Desertification de l'Afrique Tropical. p. 351, Soc. d'Edition Geographiques Maritime et Colonials, Paris, 1949, cited in Goudie,[3] 1990, p. 9.
6. Kovda, V., Land Aridization and Drought Control. p. 277, Westview Press, Boulder, CO, U.S., 1980, cited in Goudie,[3] 1990, p. 3.
7. UNEP/FAO/UNESCO/WMO, Desertification Map of the World, A Conf. 72-2, 1974.
8. UN-Desertification Conference (UNDESCON) Round-Up, Plan of Action and Recommendations, Nairobi, Kenya, 1977.

9. Dregne, H.E., Desertification in the Americas. Symposia Papers III, Desertification and Soils Policy, 12th I.S.S.S. Cong., New Delhi, 1982.
10. Sabadell, J.E., E.M. Risley, H.T. Torgenson, and B.S. Thornton, Desertification in the United States; Status and Issues, Bureau of Land Management, Dept. of the Interior, 1982, cited in Goudie,[3] 1990, p. 10.
11. Le Houerou, H.N., Desertization and climate (Translated into Arabic), Science and Society (the Arab ed. of Impact), UNESCO and Taylor & Francis Ltd., London., 166/88:79, 1992.
12. Le Houerou, H.N., The nature and causes of desertization. Arid Lands Newsletter, 3, 1, Office of Arid Land Studies, University of Arizona, Tuscon, AZ, 1976.
13. Bernus, E. and G. Savonnet, Les problems de la secheresse dans l'Afrique de l'Overt. Presence Africanne 88(4):113, 1973, Cited in Le Houerou.[13]
14. UNDESCON Secretariat, Desertification, Its Causes And Consequences, Pergamon Press, New York, Chapter 1, 1977.
15. Rozanov, B.G., Assessing, monitoring and combating desertification. 12th Int. Soil Sci. Soc. Cong. New Delhi, Symposia Papers III, p. 56, 1982.
16. FAO/UNEP/UNESCO, Provisional Methodology of Soil Degradation Assessment, p. 73, 1979.
17. UNEP, Guidelines for National Plan of Action to Combat Desertification. No. 78-2210, Nairobi, 1978.
18. UNEP, Transnational Project of Management of Livestock and Range Lands to Combat Desertification in the Sudano-Sahelian Regions. (SOLAR) A/Conf. 74/26, 1974.
19. UNEP, Transnational Green Belt in North Africa. (Morocco, Algeria, Tunisia, Libya, Egypt) A/Conf. 74/25, 1978.
20. Tolba, M.K., Can desertification be stopped? UNEP Desertification Control Bull., 2:5, 1978.
21. Bennet, H.H., Soil Conservation. McGraw-Hill, New York, Cited in Goudie.[2]
22. Balba, A.M., A Working Report on Minimum Management Program to Combat World Desertification. UNEP Consultancy, 1980, Pub. Alex. Sci. Exch., Vol. 1, 2:41 and Vol. 2, 1:23, 1981.
23. Mader, D. and M.J. Yardley, Migration, modification and merging in aeolian systems and the significance of the depositional mechanisms in Permian and Triassic dune sands of Europe and No. America, Sedimentary Geology, 43:5, 1964, in Goudie,[3] 1990, p. 55.
24. Imbaby, N.S. and M.M. Ashour, Sand Dunes in Katar Peninsula (in Arabic), Part 1, 67–92, University of Katar Documents Center, 1993.
25. Schou, A., Formation and soil characteristics of dunes. In Heathland and Sand Dunes Afforestation. FAO/SEN/TF 123, p. 37, FAO, Rome, 1974.
26. Watson, A., The control of blowing sand and mobile desert dunes. Chapter 2, p. 36, in Goudie,[2] 1990.
27. Fryreer, D.W., Principles of soil erosion control: Cultivated lands. Proc. Symp. Soil Erosion by Wind. Soil Cons. Soc. Am. and USDA-SCS, p. 16, 1987.
28. Chepil, W.S. and N.P. Woodruff, The physics of wind erosion and its control. Adv. Agron., 15:211–302, 1963, Cited in Lal.
29. Jensen, M., The aerodynamics of shelters. In FAO/DEN/TF 123:132, 1974.
30. Bhimaya, C.R., A review of sand dune stabilization and afforestation in general, with details of the practical methods of approach. In FAO/DEN/TF 123:49, 1974.
31. Lybian Min. of Agric., Pilot Project Sand Dune Fixation (Green Belt Project), Tripoli, Bull. No. 33, 1973.
32. Kerr, R.O. and J.O. Nigra, Eolian sand control, Bull. of the Amer. Assoc. of Petrol. Geologists, 36:1541, 1952, cited in Goudie,[3] 1990, p. 51.
33. Lyles, L. and R.L. Schrandt, Wind erodibility as influenced by rainfall and soil salinity. Soil Sci., 114:367, 1972.

34. Balba, A.M., Desertification in North Africa, 12th Int. Soil Sci. Soc. Cong. New Delhi, India, Symposia Papers III, pp. 14–25, 1982.
35. Finkel, H.J., The Barchans of South Peru. J. of Geology, 67:614, 1959.
36. Rosenzweg, M.L., Net primary production of terrestrial communities. Prediction from climatological data, Am. Nat. 102, 1988, cited in Goudi,[3] 1990, p. 223.
37. Coe, M., The conservation and management of semiarid rangelands and their animal resources. Chapter 8, p. 219, In Goudie.[2]
38. FAO, Agriculture Towards 2000, the 20th Session FAO C 79-24, Rome, Italy, 1979.
39. Cunningham, G.M., Total catchment management, Resources management for the future. J. Soil Cons. New So. Wales, Vol. 42, 1:4, 1986.
40. Dregne, H.E., Desertification of arid lands, 1986, cited in Goudie,[3] p. 10, 1990.
41. Office of the Arid Lands Studies, University of Arizona, U.S., News Letter, April, 1979.
42. Balba, A.M., Prognosis of salinity and alkalinity, FAO Soils Bull., 13, Rome, 1976, p. 241.
43. Balba, A.M., Sources and Protection of Soil and Water of the Mediterranean Coast of Egypt. The National report Submitted to the PAP, Split, Published in No. 6 of Advances in Soil and Water Research in Alexandria, Prof. A.M. Balba Group for Soil and Water Research, pp. 73, 1986.
44. Adu, S.U., Desertification in West Africa. Transaction of the 12th ISSS Cong. Symp. Papers III, p. 260, 1982.
45. Soil Conservation Service (SCS), Research in the Great Plains States. Misc. Pub., 902, 1963.
46. FAO Constancinesco, Soil Conservation for Developing Countries. Soils Bull., No. 30, 1967.
47. Price, M.R.S., Natural resources monitoring in Southern Africa. Report on a Technical Mission to Botswana. GEMS Mission Rep. UNEP, Nairoby, 1979.
48. FAO, Erosion by Water. FAO Agric. Dev. Report No. 81, FAO, Rome, 1965.
49. Lal, R., Water Erosion and Conservation: An Assessment of the Water Erosion Problem and the Techniques Available for Soil Conservation. p. 161–198, 1990, Chapter 6. Cited in Goudie,[3] 1990.
50. Soil Survey Staff, Soil Survey Manual, Bureau of Plant Industry, Soils and Agric. Eng. USDA. Handbook 18:251, 1951.
51. Lal, R., Soil Erosion as a Constraint to Crop Production, Chapter in Priorities for Alleviating Soil Related Constraints for Food Production in the Tropics. Int. Rice Inst. and N.Y. State College of Agric., Cornell University, p. 405, 1980.
52. Nahal, I. and K. Darmash (in Arabic), Principles of Soil Conservation, p. 303, Aleppo University Publications, Syria, 1988.
53. Ritchie, J.C., P.H. Hawkes, and J.R. McHenry, Estimating soil erosion from the redistribution of fallout 137Cs, Soil Sci. Soc. Am. Proc., 38:137, 1974.
54. Wischmeier, W.H., D.D. Smith, and R.E. Uhland, Evaluation of factors in the Soil Loss Equation, Agric. Eng., 39:458, 1958.
55. Hudson, N.W., The influence of rainfall in the mechanics of soil erosion. M.Sc. Thesis, Univ. of Cape Town, 1965. Cited in Lal,[38] 1980.
56. Fournier, F., The effects of climatic factors on soil erosion. Assoc. Inst. Hydrol. Pub. 38:6, 1956. Cited in Lal,[38] 1980.
57. FAO, Assessing soil degradation. FAO Soils Bull. 34, 1977. Cited in Lal,[38] 1980.
58. Foster, G.R. and L.J. Lane, Beyond the USLE, Advancement in soil erosion prediction. In Future Development in Soil Science Research. Soil. Sci. Soc. Am. Golden Anniversary Meeting, New Orleans, LA, p. 315, 1987.
59. Wischmeier, W.H., Estimating the soil loss equations cover and management factor for undisturbed lands, 1975, in Renard.[62]

60. Mutchler, C.K., C.E. Murphree, and R.C. McGregor, Subfactor method for computing C-factors for continuous cotton. Trans. Am. Soc. Agric. Engrs., 25:327, 1982, in Renard.[62]

61. Laflen, J.M., G.R. Foster, and C.A. Onsted, Simulation of individual-storm soil loss for modeling the impact of soil erosion on crop productivity, 1985, in Renard.[62]

62. Renard, R.G., Principles of soil erosion control: Rangelands, S.C.S. Am. Symp. Proc., p. 30, Feb. 1987.

63. Nicou, R., Contribution to the study and improvement of the porosity of sandy clay soils in the dry tropical zone. Agr. Trop., 30:325, 1974. Cited in Lal,[38] 1990.

64. Floret, C. and E. Le Floch, Agriculture and desertification in arid zones of No. Africa. UNEP/USSR Workshop on Impact of Agric. Practices, Batumi, USSR, 1984, Mim. Paper 18 pp. Cited in Lal,[38] 1990.

65. ACSAD (Arab Center for the Studies of Arid Zones and Dry Lands), Damascus, Syria, Report 17/78, 1978.

66. Armbrust, D.V., Principles of soil erosion. Detachment, movement and deposition. Proc. Symp. on Soil Erosion by Wind, Denver Co., Chapter of Soil Cons. Soc. Am. and USDA-SCS, p. 13, 1987.

67. Middleton, H.E., Wind erosion and dust storm control. Chapter 3. In Goudie,[2] 1990.

68. Fryberger, S.G., Dune forms and wind regime, In A Study of Global Sand Seas, Ed. E.D. McKee, p. 136, U.S. Geological Survey Professional Paper 1052, 1979, cited in Goudie,[3] 1990, p. 89.

69. FAO, The World Soil Charter, 21st Session of FAO Conference, Decision 8181, 1981.

70. Plan of Action for World Soils Policy, Part 2, Third Meeting, Geneva, 1982 (unpublished).

71. Garbouchev, I.P. and P.M. Ahn, Towards a soils policy. Trans. 12th I.S.S.S. Cong. Symp. Papers III, p. 69, 1982.

72. Mediterranean Action Plan, UNEP, Reports of Experts Meeting, Split, 1985 and 1987 (unpublished).

73. Rafiq, M., The Present Situation and Potential Hazard of Soil Degradation in Ten Countries of the Near East Region. FAO Regional Office for the Near East in Coop. with Soil Resources, Management and Cons. Service, Land and Water Dev. Div. and UNEP, p. 149, 1978.

74. Sant'Anna, R., The conservation and rehabilitation of African lands, a sustainable approach to national actions. 2nd African Soil Sci. Soc. Conf., Plenary lecture, pp. 17–25, 1991.

75. Kassas, M., Desertification, Environmental Education. Dosslers-All of US, 7, UNESCO Center De Catalunia, p. 1, March, 1994.

76. UNESCO, Trends in Research and in the Application of Science and Technology for Arid Zone Development, MAB Tech. Notes No. 10, p. 9–17, 1979, UNESCO, Paris.

77. Dreen, H.F., The angry sands, UNESCO Courier (Arabic ed.) No. 194, p. 14, 1977.

78. Kilinc, M.Y., Soil degradation in the Turkish Mediterranean zone. Natl. Rep. PAP, UNEP, Split, 1985.

79. Balba, A.M. (Group leader), National plan of Action to Combat Desertification in Jordan. Report ESCWA/UNEP/FAO/UNESCO Consultation Mission to Jordan, unpublished, 1986.

80. Balba, A.M., Improvement of El Wassel Rangeland and its Protection from Sand Encroachment, A Pilot Project in Oman Sultanate, UNEP Consultancy, 1982, unpublished.

81. Gerasemov, I.P., Management of the land resources in USSR. Expert meeting on World Soils Policy, UNEP/FAO/UNESCO, Rome, 1981 (unpublished).

82. Niknam, F. and B. Ahranjani, Dunes and Development in Iran. p. 21, 1975.

83. Ministry of Food and Agriculture, Govt. of Pakistan Evergreen Press, 1976.

84. Mann, H.S., Assessing, Monitoring and Combating Desertification, 12th Int. Soil Sci. Soc. Cong. New Delhi. Symposia Papers III, p. 44, 1982.

85. CAZRI, (Central Arid Zone Res. Inst.), A Case Study on Desertification, Nairobi, A/Conf., 74/11: 1–66, 1977, cited in Mann,[49] 1982.
86. Gupta, J.P. and R.K. Aggarwal, Arid Zone Research. Sci. Pub., Jodhpur, India, p. 109–114, 1988, cited in Mann,[49] 1982, p. 49.
87. Tejwani, K.G., Soils Policy in India, Need and Direction, 12th Int. Soil Sci. Soc. Cong. New Delhi, India, Symposia Papers III, p. 98–105, 1982.
88. Walls, J. (Ed.), A Report on a Seminar Sponsored by the Academy of Science of the Peoples Rep. of China and UNEP, Chapter 2, p. 8, 1982.
89. Ayoub, A.T., The status of human-induced soil degradation in Africa. Proceedings of the 2nd African Soil Sci. Soc. Plenary Lectures, pp. 9–16, 1991.
90. Brine, H., Six Thousand km in Africa under drought. UNESCO Courier (Arabic ed.) No. 176, p. 4, 1975.
91. Souissi, A., National Report of Tunisia, presented before PAP-RAC, UNEP meeting at Split, 1985.
92. Abdel-Kader, E.I., Soil degradation in the Mediterranean coastal zone of Morocco, Natl. Rep. RAC-PAP, UNEP, meeting at Split, 1985.
93. Logan, J.M., Policies for the use and conservation of soils of Australia, 12th Int. Soil. Sci. Soc. Cong. New Delhi, Symposia Papers III pp. 86–97, 1982.

INDEX

C

D

Milton Keynes UK
Ingram Content Group UK Ltd.
UKHW051949071024
449327UK00026B/2238